SUN TOWARDS
HIGH NOON

Pan Stanford Series on Renewable Energy

Series Editor

Wolfgang Palz

Pan Stanford Series on Renewable Energy
Volume 8

SUN TOWARDS HIGH NOON

Solar Power Transforming Our Energy Future

Peter F. Varadi

Series Editor
Wolfgang Palz

Contributors
Michael Eckhart
Allan R. Hoffman
Paula Mints
Bill Rever
John Wohlgemuth
Frank P. H. Wouters

PAN STANFORD PUBLISHING

Published by

Pan Stanford Publishing Pte. Ltd.
Penthouse Level, Suntec Tower 3
8 Temasek Boulevard
Singapore 038988

Email: editorial@panstanford.com
Web: www.panstanford.com

British Library Cataloguing-in-Publication Data
A catalogue record for this book is available from the British Library.

Sun towards High Noon: Solar Power Transforming Our Energy Future
Copyright © 2017 Peter F. Varadi

ISBN 978-981-4774-17-8 (Paperback)
ISBN 978-1-315-19657-2 (eBook)

Printed in the USA

Contents

Acknowledgments

My book *Sun above the Horizon* described the meteoric rise of the utilization of solar electricity (PV) from a garage operation in 1973 to 2010, when it was already a real industry. I realized that since 2010 when the price of PV modules approached the magic $1.00 per watt and even went lower, the entire solar PV business entered a new and explosive phase. New markets for PV emerged, which not only provided sustainability but also required new marketing approaches, and instead of technology, financing became the centerpiece of growth.

To complete this book, I received help from the people who were experts in some of these new areas.

I would like to express my thanks to **Wolfgang Palz**, the editor of this series, for contributing an important concluding chapter to this book about the "Outlook to the future of PV." There is no better person than Wolfgang to write about this subject, as I have the privilege to know him for more than 40 years, and I have witnessed that his forecasts were perfect. For example, when he was running the EU's renewable energy programs, he predicted, which nobody believed, that the 1995 technology and mass production will achieve module prices of $1/W PV, and as we know now, he was right.

Special thanks to **Allan R. Hoffman**, who retired from the U.S. Department of Energy a few years ago, where he served at senior positions. I had many discussions with him and he encouraged me to write a sequence to my previous book to describe PV's new phase when it became generally accepted to be one of the major energy sources for mankind and started to transform our energy system. I would like to express my thanks to Allan for the many discussions about those issues and also that when I started to establish the scope of this book, he provided me with ideas, encouragement, and also his help by writing several interesting chapters.

Michael Eckhart, in 1997, was the first to start developing and implementing new financial ideas for PV, such as the Solar Bank Initiative in South Africa and India. He was also one of the two people who, in 2013, drafted the Green Bond Principles (GBP) document; as a result, the issuance of green bonds increased immediately from $13 billion in 2013 to $36 billion in 2014 and $42 billion in 2015. Currently Michael is the managing director and global head of Environmental Finance at Citibank. I would like to thank Mike, my longtime friend, as in spite of his incredible workload, he took the time to write a section on "Financing PV" for this book.

For this book, an extremely important question was how the huge demand of these developing markets affected the supply side. Several people told me that **Paula Mints** and her SPV Market Research analyses on the PV market and the information she provides are the most accurate—second to none—and that I should ask her to write about the demand and supply in the Americas, Asia, Australia, and Europe. I have great admiration for her work. She has published many papers about the market and presented statistics with very entertaining comments. She also provided statistics for my above mentioned solar energy book. I asked her and she agreed to provide me the chapters on those continents for this book. I would like to express my thanks for those chapters.

I would like to thank **Frank Wouters**, with whom I have worked on World Bank projects, for the two chapters he contributed. The subjects of these chapters are not often discussed in PV circles and are practically unknown to the public. One is, as he calls it, "the neglected market—Africa and the Middle East." The other is about the newly formed "International Renewable Energy Agency" (IRENA). For several years until he started his consulting business, he was IRENA's deputy director-general. IRENA is an important global reach for PV.

Since 2010, huge new markets for PV have emerged. To review them, I received help from **Bill Rever**, who wrote a chapter on one of those market segments. I have known Bill since he started to work for Solarex about 35 years ago, where he continued when it was renamed BP Solar and now is consulting. His knowledge of the PV industry is the best. I would like to thank him for the chapter he wrote.

Most important for these new markets is the quality of PV modules, which is the basis for offering 20–25-year performance guarantee. **John Wohlgemuth** wrote the chapter on this topic.

John, a 40-year veteran of PV, started at Solarex—I have known him since that time—and now he is with the National Renewable Energy Laboratory (NREL). He has spent his entire career working on the quality of PV. I would like to thank him for this very important chapter.

I would like to thank Arvind Kanswal for his very professional help in preparing this manuscript for publication.

Introduction

Today, most people know what solar energy systems are and even many can spell the word "photovoltaics" and are familiar with the acronym "PV." However, the main difference between today and 50 years ago is not that people know about PV systems today, but that while 50 years ago PV was an interesting scientific curiosity, today it has become so ubiquitous and inexpensive that you can think of it as a commodity, similar to other common consumer items such as refrigerators, washing machines, air-conditioning systems, and cars. Searching "buy a PV system" on a Web search engine pops up numerous results in a split second, such as "PV System–$0 Down & Expert Installation." Numerous companies all over the World are offering their services to deliver, install, start up, and also maintain a PV system, e.g., for your house and like buying an automobile, you just have to decide whether you will pay cash, buy it on installments, or lease it. Like an automobile dealer, solar companies will even offer you financing or lease options. They also provide a 20- or 25-year warranty for the PV system, promising that it will deliver a specified amount of electricity. This warranty is much longer than what you can get when you buy an automobile or a refrigerator.

Another similarity between a PV system and an automobile or a refrigerator is that nobody is interested in how those machines or the PV system works. This book, therefore, is not going to discuss how solar systems work and their various types and technologies. A PV system is another miracle, like an iPhone and an automobile it is useful and it somehow works. It is deployed where there is light and miraculously produces electricity without fuel and without a moving part. There is, however, a difference between the solar electric PV system and an automobile. The automobile needs fuel to operate. The PV system needs no fuel and, as a matter of fact, produces fuel to operate an electric automobile. This book describes what is important to know about PV today and for the future.

This book deals with the last 7 years and discusses how we arrived where we are now. In 2010, in the entire world there were 40 GW of deployed PV systems in operation providing electric power equivalent to five fossil-fuel or nuclear power stations. By the end of 2015, the global operating PV systems provided 230 GW, the equivalent of about 30 fossil-fuel or nuclear power stations. In 2016 and in a year, it is expected that at least 50 GW capacity of new PV systems will be installed, equivalent to more than six fossil-fuel or nuclear power utilities.

By 2010, the price of electricity produced by solar systems approached the price of electricity produced by the utilities. It became much cheaper during peak demand times compared with the electricity produced by nuclear or fossil fuels even in countries with northern location, such as Germany. As a result, three gigantic new PV system markets opened up. It started with the deployment of PV systems on the roofs of private houses. Then commercial entities, such as department stores, found it profitable to install PV systems on the roofs of their buildings to produce their own electricity, replacing the electricity from utilities during the peak power time. Because of the increased demand resulting in the mass production of PV, which reduced the price of PV systems further, a third and even bigger market opened up: "utility scale PV farms." Investors realized that one can establish and sell to the utilities the large amount of PV electricity via the electric grid and make a substantial profit. The question now became not how these PV systems work but how to finance them and how the supply could keep up with the demand.

This book deals with these new gigantic utilizations of PV, issues with the development of new financing methods, and how the supply side was able to keep up with the demand. An important issue is also discussed: PV's effect on our lives and expectations for the future.

1

Meteoric Rise of PV Continues

1.1 Sun above the Horizon

The history of the direct conversion of light to electricity by photovoltaic (PV) cells started a long time ago. Ninety-five years ago, in 1921, Albert Einstein received the Nobel Prize because he deciphered the principle of the operation of the photovoltaic phenomenon, which was discovered 82 years earlier in 1839 by Edmond Becquerel. For 114 years after Becquerel, there was still practically no use for this strange phenomenon until 1953, when Daryl Chapin of the Bell Research Laboratory discovered that pure silicon wafers were able to convert light to useful amount of electricity. These PV cells—named also "solar cells"—had no important use for another five years when in 1958 an unexpected breakthrough occurred.

In 1957, Sputnik, the first artificial satellite, was put into orbit. It was operational only for 22 days when its radio transmitter's battery was exhausted. In 1958, the same happened to the US satellites SCORE and Explorer I. Their transmitters operated only for 10 and 31 days, respectively. This actually meant that the use of artificial satellites would be impractical because the required batteries needed for the supply of electricity for a long period of time for the operation of the payload would be too big and heavy.

However, when the next satellite, Vanguard I, was launched in 1958, it had one transmitter powered with a battery, and because of weight limitations several batteries could not be installed to extend its life; so in desperation an experiment was made. They installed on Vanguard I another transmitter with the same size battery they used for the first transmitter but to which a series of solar cells mounted on the outside skin of the satellite were connected. The transmitter attached to only the battery functioned for 20 days, and the other transmitter, for which the battery was connected to the solar cells and recharged by them, lasted six and a half years, when the electronic circuit of the transmitter and not the solar cells failed. It was realized that the

Sun towards High Noon: Solar Power Transforming Our Energy Future
Peter F. Varadi
Copyright © 2017 Peter F. Varadi
ISBN 978-981-4774-17-8 (Paperback), 978-1-315-19657-2 (eBook)
www.panstanford.com

utilization of PV cells was a necessity because without them the satellites were useless. Solar cells made the utilization of the artificial satellites possible. It was important that solar cells for this purpose should have high efficiency and long life under the very harsh space environment. Their price was immaterial as the cost of the PV cells was only a very small fraction of the cost of the satellites themselves and their launch, operation, and maintenance.

By 1972, the efficiency—the production of electricity—of solar cells doubled. Their reliability and life expectancy was proven. Therefore, their use for terrestrial purposes was suggested, too. The big handicap was that PV cells were extremely expensive. The price of the solar cells to be used for terrestrial purposes had to come down to 1% of the price of those used for space applications. Experts predicted that the research to achieve such a price reduction would take about 10–15 years.

Contrary to that, a few people believed that a technology totally different from the one used for the production of space-oriented PV cells was needed to produce less expensive solar cells for terrestrial purposes and that it could be developed in less than a year. That would also make possible the development of some markets for the terrestrial utilization of solar cells, which would require larger-scale production and result in declining prices. Ultimately mass production could be achieved, which would result in bringing down the price of solar cells to the level that the produced electricity would be competitive with the electricity generated by utilities. To achieve this, two companies, Solarex and Solar Power Corporation, were started in 1973. They were the beginning of the global terrestrial solar electric (PV) industry. As was described in a previous book,[1] the history of the terrestrial PV solar electricity generation industry went through phases or, as that book labels them, "Acts," as the word "*act*" is used for major distinctive sections of a story. That book describes three "Acts."

The first Act (1972–1984) started with the development of the low-cost terrestrial solar cell and module–manufacturing technology, which was accomplished in less than a year as predicted and which is still being used with modifications. Several important markets were also established requiring larger production capacity, which resulted in a gradually decreasing price

---

[1]Peter F. Varadi (2014). *Sun above the Horizon*, Pan Stanford Publishing, Singapore.

of the solar cells. During this period, the increased production of solar cells and modules resulted in a substantial reduction of its price, indicating that if mass production could be achieved, PV could become very competitive with other electricity generation systems. During that period, another equally important result was achieved: The quality of the manufactured solar cells and modules improved, and as a result their life expectancy could be guaranteed for at least 20 years. Without this quality of solar systems, which enabled them to operate for a long period—at least 20 years—the success of PV would not have been possible.

Quality is extremely important to make the PV systems "bankable," i.e., acceptable for long-term investment. Such a long-term guarantee for an industrial product is unique. How it became possible that manufacturers were able to provide such a long-term guaranty of the electricity production of PV modules is described in detail in Chapter 2.7.

The next Act (1985–1999) began when many more market areas were established, which required gradually increased scale of production of solar cells and as a result the price of solar cells was substantially reduced. However, this was also a time of unsuccessful experimentation of how to realize a market that would require and sustain the "mass production" of PV to achieve prices of the generated electricity competitive with the prices of the electricity produced by the established utilities.

The year 2000 was the beginning of the third Act, when the mass production of PV started. This was the result of the introduction of the German "Feed in Tariff" (FiT) system, which provided the incentive and sustainability for the utilization of PV systems. As was predicted, the mass production of the solar cells required no new technology. The technology developed in the 1970s as predicted resulted in sharply declining prices when the automated mass production of PV was achieved. The FiT system was basically a simple *publicly supported incentive* to deploy PV systems to produce electricity. The PV generated electricity was fed into the grid. The utilities paid for the generated kWh according to a price prescribed by the government. The price was established by the government on the basis of a calculation to guarantee that in 20 years the investor's investment in the PV system will be securely returned with a decent profit. The *Government paid nothing* and the *utilities did not pay either* because they added the FiT's mandated overpayment as a small additional cost to the electric bills of their customers like they usually add the "oil surplus" charge. As we now know, the result was astonishing. The global production of PV modules in 2000 was 250 MW and in 2009 was 8,000 MW, a 31-fold increase in the production in 10 years. The meteoric rise of the utilization of PV started.

In summary, the first two Acts of the history of the terrestrial PV industry resulted in the establishment of the technology to manufacture reliable and long-life terrestrial solar cells, modules, and systems and also the development of several market segments to be able to increase production and gradually decrease the price of the solar cells. The third Act started in 2000 with the introduction of the "Feed in Tariff" (FiT) system in Germany, which resulted in the establishment of the mass production of solar cells and modules and also in the development of automatic machinery, which was needed for the industry because of the extremely large demand of PV modules. As predicted, this resulted in a low price and the electricity generated by the PV systems was able to approach the price of the electricity generated by the conventional methods used by the electrical utilities.

The stage was now set for the next fourth Act.

1.2 Sun towards High Noon (2010 to 2016)

The year 2010 marked a new era in the terrestrial utilization of PV when it rapidly started to become a significant global electric power generation system. This could be seen from the statistics. In 2010, the global PV module production was 17,400 MW, which was more than twice what it was in 2009. In 2010, the globally deployed PV capacity was close to 40,000 MW (40 GW). Compare these numbers with those of 2015, when not the *global deployed capacity but the yearly global PV production* was close to 50,000 MW (50 GW). By the end of 2015, the world's total installed and operating PV capacity was 230,000 (230 GW), which was 5.75 times more than it was 5 years earlier and which is equivalent to the continuous (7/24) output of the total capacity of 70 nuclear power plants each of 500 MW capacity. Fresh investment in PV in 2015 was US$100 billion.

The price of solar modules decreased to a level that the capital cost for a PV electric power plant was the same or less than the plant of the same capacity operated with other fuel types except natural gas, which was still less expensive (see chapter 5.3).

The seriousness of PV's entry into the global electricity generation market can also be seen from the price of the produced electricity, which became very competitive with other generation methods. In 2015, the price of the PV-produced electricity, for example, in southern Spain was 5 US cent/kWh and in Germany 8 US cents/kWh. In Germany, the price of PV electricity in spite of the country's northern location was about the same as the cost of the electricity produced by fossil coal, nuclear, etc., which was about 15 US cents/kWh (about half of it is tax).

This means that the utilization of PV electricity generation systems became competitive with the established coal, fossil, and especially nuclear ones. A big advantage of PV over these systems is that it could also be decentralized, which means it could be used at the location where it was needed, and no transmission lines were necessary. Another advantage was that it was modular. That means that the PV system capacity can be adjusted according to the power requirements. If more power was needed, more solar modules could be added. If less electricity was needed, solar modules could be removed and used somewhere else. Compare that to a conventional (coal, nuclear, gas, hydro) centralized electricity generation system, which has one central electric power generation station, and to provide electricity to a small village the expense of the needed power line could be prohibitive. A PV system can be installed locally avoiding the cost of a power line. PV electric systems have also other huge advantages:

- non-polluting
- not contribution to global warming
- require only very little expense after installation (e.g., operating personal, maintenance, fuel transportation)
- need no infrastructure (e.g., roads, trucks, train lines, pipe line)
- have at least a 20-year guaranteed performance, which is not exposed to the availability and the fluctuation in the price of its fuel. This means the price for the generated electricity will be stable for 20 years.

PV systems' disadvantage—but also an advantage—is that the electricity generation capacity is dependent on the availability of light.

The disadvantage is that it produces electricity only during daytime and is also affected by intermittent reduction of light, for example, by a cloud casting shadow on the solar system.

The advantage, however, is that PV generates electricity during daytime and that its highest electricity production is during the hours before and after noon. The peak electric power consumption for a large number of customers coincides with this. It is also needed during the day hours, especially before and after noon. To supply this peak electricity demand, the utilities had to install so-called peaking power plants ("peakers"). Naturally, these power plants are used only when peak power is needed and therefore the produced electricity is more expensive. Utilization of PV systems, the peak power of which is produced during this time, is much more economical and therefore much cheaper. Therefore, "peakers" were actually not needed and the utilities had no income from them. As it will be shown (Chapter 5.3) the loss of this income was detrimental to the utilities.

Avoiding global warming was also an incentive to increase the utilization of renewable energy sources thereby reducing coal and fossil fuel for the generation of electricity and heat. The UN Climate Conference: 21st Conference of the Parties (COP21) in Paris sanctioned the requirement of utilizing as much as possible renewable energy systems, wind, PV, and bio. Obviously, the horrible air pollution in large cities such as Beijing, Delhi, and Los Angeles also provides an incentive to use more renewable energy sources, including PV.

By 2015, PV became competitive with coal, fossil fuel, and especially nuclear electricity generation systems. As a result, large PV capacity was deployed all over the world. How did that happen? It started with the introduction of the FiT system.

One would think that when the FiT system was established in Germany, the utilities would deploy large PV systems to take advantage of the FiT, which would provide them with the return of their investment and a decent profit for 20 years. In reality, none of the utilities purchased any PV systems. Why that happened and what the consequences were that led to drastic changes for more than 100-year-old electrical utilities is a very interesting story described in Chapter 5.3.

What happened after the FiT became law was quite unexpected. As already mentioned, the utilities did not install utility-size PV systems, but surprisingly extremely large numbers of PV systems were installed by home and farm owners. By the end of 2011, in Germany PV systems producing 24,700 MW of electricity were installed of which 81% were on private homes and barns. This story is described in Chapter 2.2.

After 2011, the trend changed. The number of PV systems for homes increased, but a great number of the large (over 1 MW) commercial (Chapter 2.4.) and later "utility-size" super large systems (Chapter 2.5) were deployed. As a matter of fact, today a 10 MW PV system is considered small because a large number of gigantic systems in the range from 100 to 500 MW were installed all over the World.

As mentioned above, in 2015 the world's total installed and operating PV capacity was 230,000 (230 GW), which required an enormous amount of investment. In 2015, in one year the investment in PV was US $100 billion. From where did all this huge capital come? An entire section, Section 3, will deal with this important issue.

2

New PV Markets Sustaining Mass Production

2.1   Utilization of the Terrestrial Solar Electricity

The electricity generated by solar modules can be utilized for two different purposes: space application, e.g., satellites, and for terrestrial application. In the first 25 years in the terrestrial PV utilization, practically all of the systems were the "standalone" applications. They are called "standalone" because the needed electricity is only provided by PV modules connected to a battery that they charge and the battery provides the electricity needed by the system continuously. In 1999, practically the entire "standalone" global PV market was about 170 MW. In 2008, it was up to 830 MW.[1]

The "standalone" is also called "off-grid," because the other possibility is to connect the solar modules directly to the electric grid, called "grid-connected" utilization. Grid-connected PV systems started to be used only at the beginning of the 1990s. In those days, PV systems' price was quite high, but the prices steadily decreased and more and more PV systems were connected to the grid. The German Feed-in Tariff (FiT) law introduced in 2000 (described in the previous chapter) was tailored for every RE system (wind, PV, bio, hydro) separately and encouraged the "grid-connected" PV systems, making it very attractive for financial reasons and caused a sustainable large demand for PV modules and ignited its wide utilization. The resulting mass production caused substantial reduction in the price, which started a "PV rush." As expected the resulting mass production made the PV modules relatively inexpensive. That combined with that the PV modules electric output was able to be guaranteed for 20 years or more (please see Chapter 2.7) resulted in an explosion in demand.

One can consider the yearly growth of the "standalone" (off-grid) market was 10% to 15%. Thus, it can be estimated that the yearly market in 2015 was about 3,000 MW (3 GW). The

---

[1]Selya Price, Robert Margolis (2008). *2008 Solar Technologies Market Report*, DOE/Go-102010-2867, January 2010.

Sun towards High Noon: Solar Power Transforming Our Energy Future
Peter F. Varadi
Copyright © 2017 Peter F. Varadi
ISBN 978-981-4774-17-8 (Paperback), 978-1-315-19657-2 (eBook)
www.panstanford.com

"off-grid" market in 15 years increased about 15 times. On the other hand, the "grid-connected" market exploded and grew about 500 times during the same period.

The combination of the low PV module prices and that it was guaranteed by the feed-in tariff (FiT) law that the utilities will buy its electric output for 20 years at a fixed price caused this astronomical increase. This astronomical increase was the result of new markets. Interestingly, the first new market was the solar roof program for residential homes, which was tried before by governments in various countries without success. The next one was when companies that had available roof surface and/or land realized that using the space to deploy PV systems will be profitable for them compared with buying electricity from utilities. Finally, investors realized that it would be a profitable business to establish utility-scale PV systems and sell the produced electricity to the utilities.

This meant that by the end of 2015 globally 230 GW (not MW) PV systems, equivalent to a very great number of conventional power plants capacity were connected to the utilities' electric grid but obviously the utilities vehemently resisted it. How was it still possible to achieve it?

It started with a political solution in the United States, which was achieved during the Carter years when the "Public Utility Regulatory Policies Act" (PURPA) was passed in November 1978 by the United States Congress as part of the National Energy Act. This law contained a provision little noticed at that time that resulted in huge unintended consequences. This provision states that utilities are required to purchase the electric power generated by private companies. According to this law, utilities must pay an equitable rate for this electricity determined by each state's Public Utility Commission. Generally, this could be the equivalent to the dollar amount the utility would have to spend to generate (but not distribute) the amount of power it receives. A similar law modeled according to PURPA was established in Germany in 1991 as part of the first "Feed-in Tariff" law for RE (e.g., wind and PV). The FiT system was later utilized in many countries, and that secured the access to the grid for RE generators in those countries.

The first result of this law was the introduction of "net metering." Net metering requires only one electric meter that can turn in both directions. The meter turns in one direction when the customer uses electricity and turns in the other direction when the customer provides electricity to the utility. This made it possible for small independent power producers (IPP) to connect small hydro or some wind and PV generators to the grid.

While PURPA was like a rickety suspension bridge that provided a connection for RE systems to the grid, assuring that customers will be paid by the utility for the electricity delivered to them, and to buy electricity from the utility if needed. The first line of defense for the utilities was not to let the RE, and especially PV-produced electricity, to be connected to their grid. Their major reason was that the RE systems connection could be improper and could cause serious problems in the operation of the grid.

PURPA'S rickety suspension bridge was transformed to a bridge with solid pillars when the Institute of Electrical and Electronics Engineers (IEEE) standard 1547 "Standard for Distributed Resources Interconnected with Electric Power Systems" under the direction of Dick DeBlasio was established and ratified in 2003.[2]

The electric grid became accessible to renewable energy-generating systems (including PV).

---

[2]Dick DeBlasio (2012). Interconnection standards for the smart grid, *IEEE 2030 Working* Group, June 28, 2012.

2.2   Solar Roofs for Residential Homes

In San Francisco on April 19, 2016, landmark legislation passed unanimously requiring that starting January 2017 all new buildings under 10 stories to be fitted with rooftop solar panels. Smaller Californian cities such as Lancaster and Sebastopol already have similar laws in place. I must confess that when we started the terrestrial PV business, we had not even dreamt that 44 years later in a metropolis like San Francisco, by law every building under 10 stories would be required to be fitted with solar panels. It is interesting to reconstruct how this happened.

We knew that the future of PV depended on generating electricity at "grid parity," meaning electricity at the price of or even cheaper than by other means. This required inexpensive PV modules and systems.

As described before, one school of PV experts in 1972–73 proposed that lots of research would be needed to find a way to make inexpensive solar cells and modules for terrestrial purposes. They estimated that it would require about 15 years to be able to find a breakthrough to be successful.

The other school was started in 1972–73, too, by four "intrepid entrepreneurs," as Richard Swanson[3] called them—Elliot Berman, Joseph Lindmayer, Peter Varadi, and Bill Yerkes—who did not believe the "prevailing view" that lots of research would be needed to develop the technology for manufacturing inexpensive Si solar cells for terrestrial purposes. They believed that a suitable technology could be developed very fast and the price of the solar modules would drop according to the produced quantity. As was described, they started the terrestrial PV industry with two factories. They found and established markets and, as predicted, the prices of PV modules decreased as the volume of production increased. When they started the "PV Industry" in 1973, the yearly global terrestrial PV production was 500 W (watt, not even one kilowatt) and the PV module price was about $50/W, and in 1989, 16 years later, as reported by Paul Maycock,[4] the global PV production was up to 33.6 MW and the price was down to $6/W.

---

[3]Richard M. Swanson (2011). The story of SunPower, in *Power for the World* (Palz W, ed), Pan Sanford Publishing, Singapore, p 532.
[4]Maycock P (ed) (2002). *PV News*, February 2002 editions.

In 1977, Wolfgang Palz became the manager of the European Commission's Development Program for Renewable Energies. He supported research but believed that PV prices would be drastically reduced if mass production could be achieved. He proposed and supported the study carried out by T. M. Bruton[5] and his group, who found that PV production using the then-existing technology and utilizing crystalline silicon wafers, by reaching the production level in a factory manufacturing 500 MW of PV modules, the price of the PV modules could be reduced to under $1.00/W. This size and even much bigger factories are in operation today and the PV module price as predicted is even much below $1.00/W and is a reality now.

The question became to find a quantum leap in PV markets to sustainably support the quantity needed for the mass production level to achieve the low price indicated in Bruton's study. It was believed that the millions of home's roofs could create a mass market and hence the mass production of PV, which would bring its prices down to yield inexpensive, close-to-grid-parity PV systems. To achieve this was one of the reasons the German, Japanese, and US governments initiated the solar roof programs.

The idea to install solar modules on the roofs of private houses came as soon as the terrestrial applications were initiated. Karl Böer, a pioneer of the terrestrial use of PV, mounted in June 1973 some modules on the roof of his house.

To demonstrate the feasibility of PV for roofs for residential application, the US Department of Energy (DOE) initiated a program to design and install PV modules on several homes. The first of these was the "Carlisle house" dedicated in May 1981 in a Boston, Massachusetts, suburb.[6] The PV system on the Carlisle house was connected to the electric grid. When the solar system generated electricity, it was fed into the grid, and when the house needed power, it received it from the grid. This, as mentioned earlier, is called "net metering" system. The US government continued with a few demonstration houses, which were all successful.

---

[5]TM Bruton et al. (1997). A study of the manufacture at 500 MW p.a. of crystalline silicon photovoltaic modules, *14th European PVSE Conference*, Barcelona, p 11.
[6]Steven J Strong (1987). *The Solar Electric House*, Sustainability Press, Massachusetts.

Interestingly, the person who first picked up the idea to install PV modules on the roofs of many houses to try to initiate the mass production of PV modules was, in 1986, a young Swiss entrepreneur, Markus Real. He called his plan "Project Megawatt," because he planned to get 333 customers to install 3 kW on the roof of each of the houses, which in total would amount to 1 MW. In 1988, with clever promotion and support from the most important Swiss newspapers, his company was able to find 333 customers to realize the project without any governmental support or subsidies. All of his customers were wealthy and able to buy the expensive PV systems. With no difficulty, within a year all of the 333 PV systems were installed and connected to the Swiss grid.[7] Real's customers were able to sell the solar electricity at the same price the utility was charging them, which was a novelty at that time. "Net metering" also reflected a breakthrough in the utility world in Europe at that time. The 333 roof-mounted PV systems indicated that the generation of electricity in decentralized PV systems and connecting them to the utility grid was feasible, but to extend such promotion on a large scale would need financial support until the mass production brought the price of PV systems down.

Governments tried to inject money for residential roof programs to achieve large-scale production of PV. In Germany, a 1,000-roof program was initiated in September 1990, which was followed with a 100,000-roof program. The 100,000-roof program was budgeted with about $700 million. It was started in 1999 and ended successfully in 2003. During those four years, 346 MW were installed.[8] For the 100,000-roof program, a 0% 10-year loan was provided by the German State owned bank, KfW (Kreditanstalt für Wiederaufbau).

In Japan as a result of the leadership of Professor Yoshihiro Hamakawa, the Ministry of Energy, Trade and Industry (METI) started in 1974 the "Sunshine project" and in 1993 embarked on the "New Sunshine Project." From 1993 to 2004, a

---

[7]Mr. Real told the author recently (September 2016) that he checked and some of the solar systems were still operational.
[8]G Stryi-Hipp (2004). Experience with the German performance-based incentive program, *Solarpower 2004*, San Francisco October 19, 2004.

total of 217,000 residential systems were installed providing a total capacity of 795.26 MW (on an average 3.66 kW/house).[9]

In the United States, in a June 1997 speech to the United Nations, President Bill Clinton proposed a Million Solar Roofs program.[10] This could have resulted in the mass production of PV, but ultimately the US Federal Government program authorized only a total of $16 million,[11] which would have been only a $16 subsidy per roof, but even that money was not spent on roofs, it was spent mostly on grants to some 90 cities.

It was realized that for two reasons these government-sponsored "roof programs" were not appropriate to bring PV module production to a level needed to establish mass production necessary to lower the prices to the $1.00/W level or lesser.

One reason was that it became clear that governments would not be able to invest the amount of money needed to achieve the mass production of terrestrial PV modules. If PV would have expected to be used by the military, such as nuclear or the semiconductors, governments would have not hesitated to invest. However, PV is a source of electricity for which, despite that nuclear electricity is expensive and dangerous, it would have been hard to convince taxpayers to invest lots of money when oil, gas, and coal were abundant. The argument for PV was that oil and gas will run out at some point and the use of oil, gas, and coal is ruining the planet we are living on; therefore, they should be replaced with energy sources that will not harm us. This was a good argument, but one could wait to spend money to replace them. It is true that oil companies received large government subsidies, but in those days there was no consensus on providing large subsidies for the small PV industry. At that time, a program to raise taxes for PV was dead at arrival in any country's legislation.

The other, and important, fault in the government programs was that they lacked sustainability for PV production. For example, the German 100,000-PV-roof program assured four years of 100 MW/year production of PV but did not assure continuity.

---

[9]http://userpage.fu-berlin.de/ffu/akumwelt/bc2006/papers/Kimura_Suzuki.pdf.

[10]The program was recommended after a meeting with the US PV Industry by Dr. Allan Hoffman, who was in 1997 the acting deputy assistant secretary for the Office of Utility Technology of the US Department of Energy.

[11]Bob Johnstone (2011). *Switching to Solar*, Prometheus Books, p 225.

What would happen after four years? Who would invest money in a production facility with that future?

It became clear that governments are not going to be able to invest the amount of money needed to achieve mass production of terrestrial PV modules and would not be able to assure sustainability. The question was whether private money would be available for PV systems.

In Europe, Wolfgang Palz of the EU started programs, and Hermann Scheer, member of the German Parliament, and Allan Hoffman of the US DOE organized conferences to find out whether private money would be available and under what conditions.

As a result, there would be practically unlimited private funds available under two conditions: (a) one could believe in the sustainability of the PV manufacturing, and (b) a reasonable return on the invested money would be guaranteed.

In Aachen, a large German city where the citizens owned the electric utility, people decided that in order to encourage the installation of PV systems on residential roofs, the utility should buy the PV-generated electricity fed into the electric grid at a price so that owners are guaranteed to get back their invested money with a reasonable profit in 20 years. The citizens of Aachen accepted the fact that because of this they would pay a little more for their electricity. This—the so-called "Aachen model" adopted on November 29, 1994—was the first useful version for PV of the first Feed-in Tariff law for Renewable Energy enacted in Germany in 1991. After the "Aachen model" (FiT) was found to be very successful, it was also adopted by many other German cities. Encouraged by this the German federal law, the "Renewable Energy Act" (Erneuerbare-Energien-Gesetz [EEG]) was initiated by Hermann Scheer and Hans-Josef Fell, both members of the German Parliament. It became a law on February 25, 2000. This law only codified the decision of the movement started with the Aachen model. It established that the German government would not provide any money or subsidy, but the law provided the sustainability of the Aachen model program for Germany.

The result of the 2000 FiT law, which was specifically tailored for each RE electricity-generating systems, such as PV, was unbelievable. In 1999, the world's total PV module production was about 200 MW. By 2003, it approached 680 MW; in three years as a result of the FiT law, it increased by 240%. By the end of 2011, the total installed PV system capacity not in the world but only in Germany increased an unbelievable 199-fold to 24,700 MW.

Surprisingly, while the FiT system would have equally benefited large and small installations, in the first years the great majority of the installed systems were the small ones installed on residential homes and in farms on barns. Table 2.1 shows the totals of the various-size PV systems installed in Germany by the end of 2011.

The German experience, as shown in Table 2.1, showed that the great majority of the installed PV system capacity was under 1 MW and totaled 20 GW. This is 81% of the total capacity installed in Germany by the end of 2011. It is quite unexpected that the great majority of this large number was in the 10 kW range and was installed mostly on private houses and farms. It is estimated that by the end of 2015, there were about 2 million residential PV installations in Germany. Their total capacity is equivalent to the capacity of several nuclear power plants.

Table 2.1   Germany: installed PV according to the size of the systems (2011)

Size of the PV system	MW installed	Percentage of the total number of 24,700 MW installed (2011)
1 MW or smaller	20,000	81
1 to 10 MW	3,050	12
10 MW and larger	1,650	7

For those who want to install their own PV roof themselves, they can find on the Internet many computer design programs for residential or other type of buildings. However, in Germany and other European countries, a large number of companies offer the planning and installation of PV systems on homes. One can easily find them on the Internet. In Germany, all of the four major electric utilities started solar divisions: RWE Solar, E.ON Solar, EnBW Solar, and Vattenfall Solar. The German utilities RWE and E.ON also offer PV installations and services to residential customers in other countries, too.

In Australia, the residential PV installations are provided by small private companies. The residential rooftop solar market in 2009 was virtually nonexistent, but as of July 2016 it exceeded

5,000 MW (5 GW) and it is estimated that over 1 million residential homes have PV installed on their roofs. It is estimated that Australian households and small businesses spent $7.8 billion in the past eight years (2007–2015) on solar rooftops.

In the United States, by the end of 2015 only 750,000 residential houses were equipped with solar power, which is low compared with Germany and Australia.

There are two reasons for this. One was that electricity was a centralized monopolistic business. Therefore, it was regulated. In the United States, the complication is that besides the federal government, the 50 states and within them the individual counties and municipalities can make decisions on matters and regulate the electrical power generation, distribution, and prices. Obviously, to install a PV system on the roof of a residential building, permits were required from one or more of these authorities. Furthermore, to connect them to the grid and sell the generated electricity to the utility—of which 3,262 existed in the United States (August 2013 figures)—was also regulated by these authorities and obviously objected to by the utilities. (This will be described in detail in Section 5.3, Chapter 5). Because of these complications, mostly local PV installers knew the regulations and the permitting process and were able to offer the installation of PV systems to homeowners. Thus, the establishment of a single electrical power law like in Germany is extremely difficult and almost impossible in the United States.

The other reason was the money required to buy and install the PV systems. Namely, cash grants and tax advantages were and are available to promote the utilization of PV to produce electricity. However, these were again different in the United States in each of the 50 States and in every localities and were also available from the federal government. None of that was, however, permanent; it was changing from time to time. The rest of the money had to be provided by the owner of the property, which means a bank had to provide it to them. In the United States, a bank similar to the German government's KfW credit facility does not exist. The loans to buy the PV systems are mostly provided by local banks. Again mostly a small local PV company could deal with this issue.

The above makes it understandable that in the United States, compared with Germany and Australia, relatively fewer residences were "solarized."

In the United States, during the last few years a very interesting and innovative business system was initiated to overcome these problems and was able to start very-large-scale deployment of solar systems on private houses.

Since 2000 when the German FiT program was started, the demand for PV modules sharply increased, which resulted in their mass production and drastically reduced prices. The global financial markets started to express big interest in its financing. At the beginning, large amounts of money were poured into financing the manufacturers of solar cells and modules globally. With solar cell prices declining, the demand for solar systems increased practically exponentially. The interesting part was that by 2010 investors looking for investment opportunities in PV were standing in line to find one. After the phase of "entrepreneurs," development of "technology," and after the FiT making possible the evolution of "mass production," PV entered the Wall Street phase, the phase of "financing" of PV. The availability of money created new financing schemes. Chapter 3 describes this in detail.

One of the new schemes was to accelerate the "solarization" of residential homes in the United States. Elon Musk—who founded the now world-renowned car manufacturer Tesla, which developed and started to manufacture all-electric automobiles— proposed a new concept never used before in the PV business. The concept was to create a company that would obtain large amounts of money from investors to be used to install PV systems free of charge on the roofs of residential houses. The company would install the PV systems free of charge and would recover its investment by selling the electricity—measured by an electric meter—to the owner and to the utility's grid at a fixed price for a guaranteed period of 20 years. The offered price of the electricity provided by the PV system on the rooftop would be more advantageous than what the homeowner's electric company would charge for the electricity as this rate would be stable for 20 years, independent from the utility's ever-changing fuel prices and inflation.

The other option the company offered was the traditional. The house owner would buy the solar installation and sell the produced electricity to the utility and that way the PV system would pay for itself in less than 20 years.

Based on Elon Musk's concept, his cousins, Peter and Lyndon Rive, in July 2006 founded California SolarCity Corporation, which was able to raise from investors approximately $2.8 billion investment. SolarCity selected from the above-described patchwork of regulations, permits, possibility of "net metering," grants and tax advantages the locations in the United States where the most favorable conditions existed for the installation of PV systems and developed in those areas its sales and installation centers. The concept's success can be judged from the fact that at the end by 2015, only six years after it was founded, SolarCity received an order in every 5 minutes of a 24 hour day, and the company had 13,000 employees.[12]

The success of the concept also shows that by now more companies have entered the field and they also were able to raise lots of money from investors and began using SolarCity's concept to deploy PV systems on the roof of residential homes.

The three most prominent of these companies

SUNGEVITY was founded in 2007 in California and raised approximately of $850 million investment. In 2014, it expanded to Europe and founded Sungevity Netherland.

SUNRUN was founded in 2007 in California. It received approximately $970 million.

VIVINT SOLAR was founded in 2011 in Utah. So far, the company received approximately $850 million investments. SolarEdison was scheduled to buy VIVINT in 2016, but the deal was not completed because SolarEdison went bankrupt in April 2016.

The most unbelievable thing happened in the spring of 2013 when the largest German utility Rheinisch-Westfälisches Elektrizitätswerk AG (RWE) realized that the already-existing close to two million homes and farms equipped with PV systems on their roofs feeding their electricity to the grid represent a huge market for an electricity storage system designed to make them more independent from the utility.

This is quite an unbelievable development if you think about it. However, the advantage for the utilities is that this leads to less PV systems feeding electricity into the grid at peak power time. The disadvantage will be that they lose a customer as these mini-utility owners may want to become fully independent from the utility, especially as they were able to utilize the German government's subsidy to buy the electricity storage system.

---

[12]Tesla Factory, an electric automobile manufacturing company, is acquiring SolarCity at the end of 2016.

On the other hand, RWE obviously realized these customers will be lost for the utilities but by offering to make them independent will convert them into a different customer—namely customers of RWE's "HomePower solar" system. On its Web site RWE carried an advertisement explaining to their electricity customers the advantages of becoming independent from the electric utility: *"What you should have is a **solar electricity storage** that makes it possible to utilize the solar electricity exactly when you need it. This independence offers you as of now the **RWE HomePower solar."** [13]

This example illustrates that the meteoric rise of the utilization of PV in Germany had some visible effect on the German electrical utility system. The utilities ignored for 40 years the existence of these experimental gadgets called PV. Then in 2012 the four large German utilities suddenly realized that PV is changing their entire business model. First, they all turned "green." A year later, they realized that to make their electricity customers independent from them would be beneficial for them and also for the home owner. The utilities will lose a very small customer but open up a new large business for themselves. The homeowners will receive a stable electricity price for 20 years, which was lower than what they would have bought from the utility.

The decentralized PV systems' effect on the future of the 100-year-old German electric utilities is discussed in Chapter 5.1.

With the advent of the all-electric cars, the price and size of electricity storage systems have drastically reduced (This will be discussed in Chapter 2.8). This made possible the large-scale combination of PV with electricity storage. Utilization of PV in combination with electricity storage (e.g., batteries) to make the homes independent was also started and opened up the automation and modernization of the home's electric system.

It was realized that more and more of today's home appliances require not only AC, but some require DC electricity, too. The large German company Bosch realized this in 2013 and

---

[13]http://www.rwe.com/web/cms/mediablob/de/2005380/data/1108176/3/ rwe-effizienz-gmbh/ueber-rwe-effizienz/praesentationen-und-broschueren/ Broschuere-Die-Basis-fuer-mehr-Unabhaengigkeit.pdf (the advertisement is in German).

developed the "Smart PV System" (BPT-S 5), an "electronic brain" to coordinate the electricity system in the home. The "Smart PV System" is a combination of a PV system with electricity storage and an inverter (used to convert DC produced by the solar module to AC, the type of electricity distributed by the utilities to customers) and an "electronic brain" distributing the electricity, DC and AC, to where it is needed. The use of the "Smart PV System" was recently started and it is envisioned that this new PV electricity storage and system controller will be the largest selling product in the future. It will be used not only for small residential installations but also for larger ones for businesses, such as department stores. Figure 2.1 shows a schematic of a future "Smart PV system" controller.

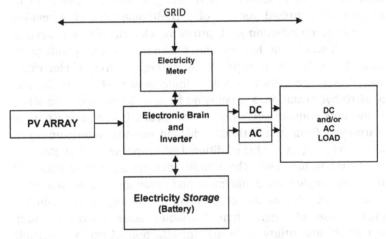

Figure 2.1   Schematic of the "Smart PV System" for homes.

The Smart PV system electronic brain supplies DC from the PV array directly to charge the battery. It provides from the battery DC to the appliances operated on DC. With a built-in inverter, the "brain" supplies electricity to the appliances operating on AC. If the battery and the appliances need no power, it sells the PV system–generated electricity through the inverter as AC through a meter to the grid and buys electricity from the grid and converts it to DC for the battery if the solar array does not produce enough electricity for the load. Obviously, the "electronic brain" can be controlled remotely using an iPhone.

2.3 Grids, Mini-Grids, and Community Solar

Allan R. Hoffman[14]

A number of factors have come together in recent years that are beginning to shape our future electrical energy systems. One major factor was Hurricane Sandy in the US Northeast, a Category 3 hurricane that lasted from October 22 to November 2, 2012, and killed 285 people. It had a lasting effect on local officials, who had to deal with the loss of power, heat, and clean water that affected millions of people in more than a dozen states, and who were determined not to let that happen again. One important response was to turn to increased use of mini-grids, which are discussed below in some detail. A mini-grid, which can be thought of as a miniaturized version of a traditional grid, is a smaller, low-voltage distribution grid, providing electricity to a collection or a community of houses and businesses, often a village or small town. It can be supplied by a single source of electricity, e.g., diesel generators, a solar photovoltaics (PV) installation, a micro-hydro station, a small wind farm, or a hybrid combination. It includes control capability, which means it can connect to or disconnect from a traditional grid and operate autonomously if necessary—e.g., when the traditional grid experiences an outage.

Other factors were the steadily decreasing cost of solar PV, the increasingly wide deployment and resultant greater familiarity with PV, the decreasing cost of energy storage that facilitates greater use of intermittent (variable) solar energy at both household and utility scale, the introduction of smart electronic grid technologies that allow optimized load control, maximum grid efficiency and the ability to island (separate a mini-grid from the larger grid), the availability of energy-efficient lighting and appliances to limit and even decrease grid demand, access to more favorable financing schemes for solar and energy storage projects, and a more favorable policy environment for solar deployments and mini-grid applications.

In general, grids are collections of wires, switches, transformers, substations, and related equipment that enable the delivery of electrical energy from a generator to a consumer of that energy. Most people are familiar with elevated wires strung

[14]Allan R. Hoffman's biography is on page 292.

between grid poles in our cities and byways. The first grid, for delivery of alternating current (AC) electricity, was put into operation in 1886. For more than a century, the traditional grid has been a one-way distribution network that delivers power from large centralized power generating stations to customers via a radial network of wires. Regional grids, when integrated, constitute a national grid, something the United States and other national electric utility systems are still trying to achieve. Transmission lines are long-distance carriers of electrical energy transmitted at high voltages and low currents to minimize electrical losses due to heating in wires. This high voltage is then reduced via transformers to lower voltage, usually 120 or 240 volts, to supply local distribution networks that bring the energy to our homes and businesses. The US Energy Information Administration estimates that national electricity transmission and distribution (T&D) losses average about 6% of the electricity that is transmitted and distributed in the United States each year.

While the traditional grid has brought the benefits of electricity to billions of people for many decades, its shortcomings have become more visible in recent years. As many weather events have demonstrated, and Hurricane Sandy emphasized, the traditional, centralized grid is vulnerable to disruption by extreme weather events as only a small fraction of T&D wires are underground. The utility industry has usually, but not always, resisted burying their wires because of high costs, but is putting increased effort into trimming trees that can fall on and disrupt power lines. Traditional grids are also vulnerable to physical attack, cyber-attack, and even large sun-powered solar storms that strike Earth occasionally and interact with the T&D system acting as giant antennas.

Grid systems with sensors and computer controls are referred to as smart grids. They can gather and analyze analog or digital information on the behavior of suppliers and consumers, and then use automation "to improve the efficiency, reliability, economics, and sustainability of the production and distribution of electricity." The critical issue of vulnerability of such smart systems to cyber-attacks has only begun to receive careful attention in recent years as the hacking phenomenon has surged and the ability to interrupt remote industrial activities via computer viruses such as Stuxnet has been demonstrated.

To protect against this vulnerability, considerable effort is going into developing software that is resistant to hacking, but this is proving difficult to achieve. Another approach is to move away from the historic centralized grid and move to a grid system where disturbances can be isolated once detected and thus do not affect other parts of the grid. This will require distributed generation sources that supply unaffected parts of the grid, and could be other centralized generators that can be tapped or local renewable energy sources (wind, solar) that are not in the disturbed grid sector.

Traditional grids are expensive, and extending these grids from urban to remote areas often cannot be justified economically. This is particularly true in developing countries where most of the world's 1.5 billion people without access to electricity reside. Improving access to modern energy services in rural areas is a major development priority, and use of mini-grids is of particular interest in developing countries that lack traditional grids and the financial means to construct them.

In fact, mini-grids may be the only way to bring electrically powered energy services to these populations. A 2013 workshop, "Mini-Grids: Opportunities for Rural development in Africa," organized by the Africa-EU Renewable Energy Cooperation Program (RECP) in Tanzania focused on this rapidly emerging option. As stated in the workshop description: "Given Africa's abundance of renewable energy resources, the widespread existence of isolated, expensive, highly-subsidized fossil-fuel based mini-grids on the continent, very low grid connection rates, the often low levels of electricity demand from households, the high costs associated with grid extension, the lack of reliable, centralized generation capacity and increasing levels of densification as a result of ongoing urbanization, renewable energy and hybrid-based mini-grids provide a practical, efficient energy access solution." The use of renewables reduces fossil-fuel use and their costs, reduces carbon emissions and other environmental impacts, and helps create local jobs and economic development.

A typical mini-grid delivers AC to its consumers. Another type of mini-grid that delivers DC, which is provided directly by PV before it is converted to AC, is usually called a micro-grid. Still another variation is the skinny-grid, which emphasizes the use of energy-efficient light bulbs and appliances to reduce consumer demand and thus allow the use of thinner and less expensive connecting wires between generators and end users.

As the technology for implementing mini-grids is available today, the critical decision factor for most municipalities in developing and developed countries considering a mini-grid is how to handle initial investment and on-going costs. As previously mentioned, a mini-grid can be powered by a wide variety of energy resources, but most attention today is on mini-grids with solar PV generation combined with electrical energy storage. Such mini-grids are referred to as Community (or Shared) Solar, and are discussed below. Today typical mini-grid facilities have installed generating capacities between 5 and 300 kW, but larger systems up to a few megawatts do exist. They offer the following advantages:

- They can be installed more quickly than a traditional centralized grid.
- They have the flexibility to upgrade their capacity to meet a community's increased energy needs.
- They can interface with regional or national grids or operate autonomously when these larger grids are down.
- They reduce power theft, which is more common on centralized grid systems.
- Reliability of supply is often greater from hybrid mini-grid systems when compared to grids dependent on a single generating technology.
- Hybrid mini-grid systems, which often incorporate a 75–99% renewable energy supply, reduce carbon emissions and line loses due to proximity to load.
- They can provide emergency power to larger grids with which they are interconnected as well as black-start power to restart the grid after an outage.
- They can reduce electricity costs over time as fossil fuel costs increase and renewable energy costs decrease.
- Mini-grids of appropriate scale may prove to be an attractive option (a "sweet spot") for utilities as their business models are required to change under the increasing pressures of climate change, distributed generation, and reduced costs of renewable energy and storage.

This is not to say that successful deployment of mini-grids does not present challenges as well. Some critical needs are the accurate assessment of

- population density within the boundaries of the mini-grid, which impacts load and project economics;
- geographic constraints, which will impact the cost of infrastructure, imported fuel cost if diesel is part of the system, and general operation and maintenance costs.
- adequacy of primary renewable energy resource (solar, wind, hydro, biomass) and seasonal resource fluctuations;
- requirements for training of local personnel to install, operate, and maintain mini-grid systems;
- friendliness and predictability of the local regulatory environment;
- the need for attention to issues of market access, small business development, and working with local financing institutions.

How to finance a mini-grid often presents the greatest challenge. Four different models of mini-grid financing have evolved: a community-based model in which the community becomes the owner and operator of the system and provides maintenance, tariff collection, and management services; a private sector model in which a private entity creates and operates the mini-grid; a utility-based model in which a utility takes responsibility for all or part of the mini-grid; and a hybrid business model that combines aspects of the three previous models. The utility-based model is widely used for rural electrification in developing countries where strong utilities exist. Where strong utilities or private sector entities do not exist, the community-based model is commonly employed.

Each model has its variations depending on local conditions and available resources. To date most mini-grid schemes use diesel or small hydro power generation and are operated and maintained by governments. While the overall level of mini-grid development still remains low in most developing countries, a few countries have moved ahead. China has an estimated 60,000 mini-grids, and Nepal, India, Vietnam, and Sri Lanka have an estimated 100–1,000 each. These numbers should increase rapidly as renewable energy, particularly solar, replaces diesel fuel and becomes a dominant feature of new mini-grids in both developing and developed countries. One example is Australia, which has excellent solar, wind, and ocean energy resources, where electricity markets are anticipating a major change, and mini-grids are seen as a solution to Australia's rapidly increasing electricity costs.

While all four financing models have been employed in various projects, the most common to date has been the community model. However, as more experience with renewable energy and mini-grids is gathered, utilities adjust their business models to the new realities in the energy sector, and private companies explore the new financial opportunities, we can anticipate greater use of the other three financial models.

Community Solar: The early adoption of the community model came about because it addresses an important need as solar energy goes mainstream: How does one participate in the solar revolution if you can't effectively put solar on your roof due to house orientation, shading by trees and other houses, are restricted from placing solar panels on your roof by cluster or other regulations, are restricted from placing solar on your business roof by other building-support equipment already being there, live in a condo in a multistory building, are a renter, or have limited means to finance a solar installation? Nearly three quarters of American households fall into one of these categories. Community solar, a mini-grid powered by solar PV that allows a few to many energy consumers to share the benefits of one local solar "farm" (sometimes called a solar garden), offers a solution to this conundrum. It allows these "excluded" consumers to receive electricity from the mini-grid without having to install the solar panels on their own roofs. However, it comes with its own complexities arising from the need for power purchase agreements, limited private partnerships, special purpose legal entities, varying regulatory environments, and the fact that each utility, public or private, has its own interests. This creates a confusing market situation for solar PV vendors, project developers, banks, and other financing bodies. Nevertheless, the deployment of community solar is happening and happening quickly in many locations. In the United States, it started in 2006 with the first shared solar energy project in Ellensburg, Washington, which cost more than $1 million to complete and placed solar panels on city-owned ground near soccer and baseball fields. It would grow to 109 kW over the coming years, enough to power a dozen homes.

Progress since then has been rapid, and according to SEIA, the Solar Energy Industries Association:

- There are 25 states with at least one community solar project on-line, with 91 projects and 102 cumulative megawatts installed through early 2016.
- At least 12 states and D.C. have recognized the benefits of shared renewables by encouraging their growth through policy and programs.
- Four states—California, Colorado, Massachusetts, and Minnesota—are expected to install the majority of community solar over the next two years.
- The next five years will see the US community solar market add an impressive 1.8 GW, compared to just 66 MW through the end of 2014.

What is also true is that while solar PV has been growing rapidly, driven in part by significant drops in cost (almost 70% from 2005 to 2015), solar PV accounted for only 1% of US electricity in 2015. One obvious conclusion is that community solar, defined as mid-sized projects in the range 0.5–5 MW, is just beginning to grow and represents a large untapped market. This market has been estimated to be as large as 30 GW by 2020 if one includes solar power owned or purchased by utilities. Rocky Mountain Institute (RMI) estimates that the long-term market potential for this latter category of community solar, often referred to as community-scale solar, is more than 750 GW.

One final word on the role of smart grids in facilitating the integration of renewable energy into the grid. Despite the variable nature of solar and wind energy, by using the control features of increasingly sophisticated smart grids and the use of energy storage, this integration can be done safely and cost effectively with high levels of renewables' penetration. The International Renewable Energy Agency (IRENA), headquartered in Abu Dhabi, has addressed this issue in a comprehensive November 2013 report entitled "Smart Grids and Renewables." It acknowledges that "much of what is known or discussed about smart grids and renewables in the literature is still at the conceptual/visionary stage..." but includes "...several case studies that involve actual, real-world installation and use of smart-grid technologies that enable renewables." As with solar PV, mini-grids, and community solar, smart grid technology is just beginning to emerge and will be an important factor in shaping our future electrical energy systems.

2.4 Commercial PV Systems

As mentioned in a previous chapter (2.2) it was very surprising, that while large and small installations would have equally benefited from the German FiT program, until the end of 2011 the great majority, over 80% of the total installed electrical capacity, were the small PV systems on residential homes and in farms on barns and only about 20% by commercial enterprises in Germany.

Since 2010 the table was turning. More and more and bigger and bigger PV systems were deployed by or for commercial companies. These PV systems are installed to produce electricity for the commercial entities'[15] self-use and only the excess is being fed into the grid to be sold to utilities.

At first commercial entities installed PV systems to indicate their environmental stance. After 2010 when the cost of electricity generated by the PV systems became not only competitive but also cheaper than buying from utilities, commercial users started to realize that besides showing their environmental attitude, they could profit from utilizing PV-generated electricity. Besides financial benefit, they would also be able to get their electricity for a fixed price for 20 years, and by installing electricity storage their facility would not be exposed to occasional power failures.

Figure 2.2   Solarex building, 200 kW multicrystalline Si roof, Frederick, Maryland—1982.

---

[15]The Solar Energy Industry Association (SEIA) publishes a yearly update of the U.S. Commercial Solar Users (http://www.seia.org/research-resources/solar-means-business-2015-top-us-corporate-solar-users).

The first was a 200 kW PV system deployed in 1982 over the roof of the solar cell and module manufacturer Solarex Corporation's building (Fig. 2.2). It was to demonstrate that for commercial purposes a large (at that time) PV system can be utilized to provide electricity, in this case primarily to the computer and the control systems of the production machinery. It was an "off-grid" system. The PV system was charging a battery bank, which through an inverter supplied AC for those important equipment. It eliminated the problem of occasional power failures of the electricity provided by the utility. The system operated for about 30 years when the building on which it was deployed was demolished. Shortly after Solarex's PV system was completed, an even bigger 300 kW roof-mounted system was installed in 1984 on Georgetown University Intercultural Center[16] (Washington DC). The purpose of this PV system was to indicate the environmental stance of the donor.

In Europe between 1982 and 1990, the European Union sponsored for demonstration purposes the installation of several small (the largest was 50 kW) "standalone" PV systems for commercial use. Joachim Benemann, president of Flachglas Solar GmbH (later renamed to "Flagsol"), developed the building-integrated PV system (BIPV).[17] In the BIPV system, the solar modules were embedded in the walls and/or roof of the buildings. The first of these (4.2. kW system) was installed in 1991 on the southern façade of the city of Aachen's (Germany) electric utility's (STAWAG) headquarters building. Most of these BIPV buildings were built to express the environmental stance of the owner of the building.

The most spectacular building of that period was the "Academy for Continuing Education" in Herne, near Gelsenkirchen, Germany[18] built in 1999. The specially built PV modules forming the roof and the south wall of the building produced 1 MW electricity. The system is still in operation.

The picture of the building was taken from above. It can be seen that the roof was covered by PV modules, but there were areas that had clear glass. The reason for this was that the building

---

[16]http://maps.georgetown.edu/interculturalcenter/.

[17] Peter F Varadi (2014). *Sun above the Horizon*, Pan Stanford Publishing, Singapore.

[18] All the information about the Academy Mont-Cernis Herne was confirmed by Ms. U. Martin, Building Management, Herne.

walls and roof was like a big 176 m (574.4 ft.) long, 72 m (236.2 ft.) wide, and 16 m (52.5 ft.) high glass box inside which the climate is maintained similar to that of Nice (France), the city on the Mediterranean.

As one can see from Fig. 2.4, inside the big glass box were several buildings and a garden. The roof segments where the PV modules were located were above the buildings, they absorbed the sunshine to produce electricity and not letting the sunshine through to warm up the roof of the buildings. On the other hand, the transparent area where no PV modules were located was over the garden, so the plants could get sufficient amount of light.

Figure 2.3 Roof of the Academy for Continuing Education, Herne, near Gelsenkirchen.

By 2010, when the demand fueled by small PV systems led to the mass production of PV modules and resulted in price drop from above $3/W to little above $1/W, commercial entities' interest picked up. One of the first was the roof-mounted system on a fruit and vegetable distribution center in Perpignan in the southern part of France (Fig. 2.5).

Figure 2.4   Inside the Academy for Continuing Education.

Figure 2.5   Perpignan, France. Solaire France "Sunstyle" PV roofing tiles (8.8 MW).

In 2011, it was the world's largest PV roofing system. The installed 97,000 PV roof tiles produced 8.8 MW. This system produces annually 9,800 MWh of solar electricity, which corresponds to about 10% of the electricity consumption of the city of Perpignan (population 118,000). The system was realized with the innovative PV roofing product SUNSTYLE, developed by Solaire France. The PV roof tiles are manufactured in partnership with the French company Saint Gobain Solar.

In Australia, Clean Energy Finance Corporation (CEFC), an Australian government-owned *Green Bank*, was established in 2012 to financially support Australian businesses to invest in solar. It was reported[19] that by now 23% of businesses generate some portion of their electricity supply using solar PV.

In Singapore, starting in January 2016, solar energy developer Sunseap Group will provide Apple with 100% renewable electricity from its portfolio of solar energy systems built atop more than 800 buildings.[20]

In the United States, more and more large or small commercial business presidents and their boards—except of the oil and coal producers—believed that global warming is a threat to mankind and made plans to convert their source of electricity to RE (e.g., PV), especially if it would be beneficial to the bottom line of their company's financial statement. That is why when it was realized that PV electricity became less expensive than buying it from the utilities, America's largest companies and many other smaller businesses started to use more and more PV systems for their electricity needs.

Commercial PV electric systems in the United States

By 2010, the price of PV electricity became very competitive with the electricity produced by the utilities with other generating methods such as fossil fuel. Especially it was the case that utilities as described in Section 1.2 had to install extra power plants ("peakers") for the daytime peak electricity demand. This "peakers" operate only during the peak demand time; therefore the peak demand electricity price is more expensive. PV systems, however, produce the most electricity during this time and obviously it is much cheaper than buying it from the utilities. Furthermore, the price of PV electricity can be guaranteed for 20 years and contrary to the utilities it uses no fuel and therefore it cost is independent from inflation and the fluctuation of availability. In addition, the use of PV instead of fossil fuels is advisable to slow down global warming.

---

[19]http://reneweconomy.com.au/2016/one-quarter-of-australian-businesses-generates-solar-power-survey-says-57289.

[20]http://www.reuters.com/article/apple-singapore-energy-idUSL1N138304-20151115.

As a result, it became clear that there was no need to wait for a miracle until "some technology breakthroughs would happen"—as stated by some people, including the CEO of a big oil company—and the utilization of PV for electricity production for commercial PV systems skyrocketed in the United States. The only question remained how to finance to become green.

This was solved when the financial community realized that PV became a large and serious new financial market and therefore various financing systems were established. Section 3, "Financing," provides details of the various financial arrangements developed to finance commercial PV systems. Chapters 3.4.2 and 3.5 are specifically focused on the financing of commercial PV systems.

Commercial users of electricity had two approaches to obtain the cheaper green electricity: PV and wind. They could obtain it either from a remote facility of a "green electricity" producer by signing a Power Purchase Agreement (PPA) or from PV systems deployed on their own facility and/or land.

It is reported[21] that Google was the first company in 2012 which started to contract to use "green electricity" (0.05 GW) signing a PPA, but in 2015 already 19 companies had purchased 3.21 GW from independent power producers (IPP).

It is, however, very interesting that many commercial organizations started to use their own property, building roof and/or land, to deploy solar on a very large scale. In this case the company either invested its own money or an investor built the PV system and provided the electricity for a long term (20 or 25 years) and fixed price PPA to the company.

This is a simple description of the two possibilities the commercial company can select to take advantage of the cost advantages of a PV system. The reality is somewhat more complicated as there are tax advantages. These are discussed in detail in Section 3.

The US Solar Energy Industries Association (SEIA) provides a yearly report entitled "Top U.S. Corporate Solar Users."[22] The data presented in SEIA's 2015 report indicate that the annually installed commercial PV capacity in 2010 was about 300 MW and one year later in 2011 increased 2.7 times to a little more than

[21]Rocky Mountain Institute, Jennifer Runyon, *Renewable Energy World*, July 8, 2016.
[22]http://www.seia.org/research-resources/solar-means-business-2015-top-us-corporate-solar-users.

800 MW and another year later in 2012 to 3.7 times to about 1,100 MW (1.1 GW). After that the yearly additional installed PV capacity stayed around this level during the next years and this is estimated also for 2016.

This chapter will only detail the commercial PV systems that are deployed on the property of a commercial company, the electricity of which is utilized for commercial purposes by the company.

SEIA's 2015 report indicates that the top 25 US companies installed 1,462 individual PV systems on site. Walmart installed the most 348 PV systems. Besides Walmart, seven other department and grocery stores are among the top 25. One of these stores is IKEA, which is the most geographically expansive; it has 42 solar-powered stores in 22 states in the United States.

The commercial PV systems are mounted either on the roof of the commercial buildings or on the land belonging to the company or on the carports near the building(s).

Rooftops are the majority—more than 80%—of the commercial PV installation systems. An example is the IKEA store in Atlanta, Georgia. The top view of the facility with PV system mounted on its roof is shown in Fig. 2.6. The 1.035 MW system started its operation on July 23, 2012. IKEA Atlanta's system will produce annually approximately 1,416,502 kWh of clean electricity.

Figure 2.6   IKEA store, Atlanta, Georgia (courtesy of IKEA).

Another example is IKEA's PV roof-mounted 1.46 MW PV system made of crystalline Si modules in Memphis, TN (Fig. 2.7). The PV system starts its operation by the end of 2016.

The largest rooftop mounted solar PV system in the United States until recently was the 5.38 MW one on Toys R Us' largest distribution center in Flanders, New Jersey.[23] The system was dedicated on August 11, 2011. Toys R Us signed a PPA with Constellation Energy (now Exelon), which built the system and is operating it.

Figure 2.7   IKEA store, Memphis TN (Courtesy of IKEA).

At present, the largest US rooftop commercial PV system is the 8.3 MW MGM Resorts International's Mandalay Bay Convention Center in Las Vegas, Nevada. The installation was completed on July 7, 2016. NRG Energy, Inc., is the owner of the array and sells the electricity to MGM Resorts International on a 25-year PPA. The system has the capacity to power 25% of the entire electricity demand of the Mandalay Bay Resort and Casino complex.

Ground-mounted PV systems were installed, for example, by GM, Verizon, and L'Oreal for their manufacturing and data centers.

---

[23]https://www.toysrusinc.com/press/toysrus-inc-powers-up-north-americas-largest-rooftop-solar#sthash.ApKd0Gji.dpu.

Carports system sizes vary from a few kilowatts to over 1 MW. One example on the lower end is a 7 kW carport canopy for an office building in Budapest, Hungary (see Fig. 2.8).

Figure 2.8   Budapest, 18 Márvány Street (courtesy of Valéria Széll).

Figure 2.9   1.3 MW solar parking canopy (courtesy of Solair Generation[24]).

---

[24]http://solairegeneration.com/project/connell-company/.

Another example on the high end is a 1.3 MW solar parking canopy with electric car charging stations installed at L'Oréal USA's second headquarters in Berkeley Heights, New Jersey. Five-hundred fifty employees enjoy the benefits of the covered parking. These arrays generate enough clean energy to cover 60% of the site's annual electricity needs (more than 1.6 million kWh per year).

Rooftop, ground-mounted and carport PV systems

Commercial users are selecting one or the other types of systems, but there are commercial users that use two or all of the three types for their PV systems. L'Oreal USA has PV systems mounted on roofs, on the ground, and on carports. It started its commitment to solar in 2011 with the installation of a 1.4 MW PV system at its Piscataway, New Jersey, manufacturing facility. By the end of 2013, it had installed more than 10.703 MW of photovoltaic systems. Out of this 81% was rooftop, 7% was ground-mounted, and 12% was on carports. All the other commercial PV systems in the United States have about the same percentages of the three different mounting possibilities being used.

The growth of the commercial utilization of PV electricity one can be seen by L'Oreal's example. The first PV installation was in 2011—a 1.4 MW PV system—and 5 years later by the end of 2016, it will have a total of 13.5 MW capacity[25] of PV in operation.

Apple has to be mentioned for two reasons. First, it planned its entire global operation to be 100% RE powered and it is far on the way to achieve it. Second, it formed a subsidiary Apple Energy LLC to sell its surplus PV electricity not only at wholesale prices to the utilities but also on a retail level (Chapter 5.3) to individual customers.

In order for Apple's operations to be globally 100% RE powered, it was betting on solar PV-generated electricity. Google, which also planned to operate using RE electricity, is betting more on wind.

Apple started to rely on its own solar PV electricity some time ago. In North Carolina, a 20 MW solar PV system in 2012 was the first followed by the deployment of two more 20 MW PV systems also in North Carolina. The electricity is being used by its nearby

---

[25]http://solarindustrymag.com/loreal-usa-turns-to-solar-to-exceed-environmental-goal.

data center. Apple is establishing another data center in Nevada and is planning to run it with PV electricity. For this reason, it is planning to establish a 70 MW solar PV field near the data center.

In 2015, it invested $850 million in a 150 MW PV solar farm built by the Arizona-based First Solar Corporation. Apple will receive 130 MW of the solar field's electricity production for 20 years. The deal will supply enough electricity to power all of Apple's California stores, offices, headquarters, and a data center.

Apple is building its Campus 2 in Cupertino, California. The site will be powered by 100% renewable energy, making it one of the most energy-efficient buildings in the world. The solar panels installed on the roof of the campus will generate 16 MW of power, sufficient to power 75% during peak daytime.

Considering all of its PV systems producing electricity, it was natural to start Apple Energy LLC to be able to sell surplus electricity not only to a utility but also to end users as they pay much more for it. Apple's application to the US Federal Energy Regulatory Agency (FERC) to be able to sell retail electricity originating from its PV system was approved.

Commercial businesses relying more and more on PV electricity are making a big step to stabilize their expenses by turning electricity from a variable to a fixed expense and in the process becoming an independent power producer (IPP).

2.5   Utility-Scale Solar

Bill Rever[26]

The dream of a world fueled by ubiquitous solar power has long had competing and distinct visions. Would this utopia consist of solar panels mounted on our homes, offices, factories, schools, etc., making them all self-sufficient in energy? As explained in chapters 2.2.and 2.3, this is now becoming reality. Other visions of this utopia are that the future would be powered by vast solar arrays in remote deserts or by huge satellites capturing solar energy, converting it to electricity and transmitting that to Earth. While the appeal of small-scale solar generating power at the point of use is strong and distributed solar generation is a very important part of the mix, it is not sufficient to achieve a future fully powered by solar energy. An analysis of energy data for the United States clearly bears this out. A recent National Renewable Energy Laboratory (NREL) report[27] calculated the total national technical potential of rooftop PV at 1,118 GW of installed capacity and 1,432 TWh of annual energy generation, equal to 39% of US national electricity sales in 2013. Yet, electricity only represented 39% of US primary energy consumption in 2014.[28] So viewed in the context of the overall energy picture, rooftop solar could at best contribute only 16% of our total energy needs under current conditions. While no global calculation like this exists, there is every reason to believe it would be similar to that for the United States.

Technological improvement will likely improve the rooftop potential over time, but clearly if the world is to shift to solar for the majority of its energy, rooftops are only part of the answer.

The use of space solar power for terrestrial energy needs has not progressed beyond the stage of small Earth-based experiments and so is not considered likely to become reality

---

[26]Bill Rever's biography is on page 293.

[27]Pieter Gagnon, Robert Margolis, Jennifer Melius, Caleb Phillips, Ryan Elmore, (January 2016). *Rooftop Solar Photovoltaic Technical Potential in the United States: A Detailed Assessment*, National Renewable Energy Laboratory, Technical Report NREL/TP-6A20-65298, vii.

[28]*U.S. Energy Facts Explained*, http://www.eia.gov/energyexplained/index. cfm?page=us_energy_home (last updated: March 26, 2015).

anytime soon. In contrast, there has been a dramatic price reduction in electricity generated by large, so-called "utility scale" solar PV systems, which have now become competitive with conventional fossil or nuclear-fueled central stations plants and are a significant fraction of the new electricity generation installed globally.

Utility-scale systems fall into the broad category of "grid-connected" solar as distinguished from "standalone" systems as described in Chapter 2.1. The difference between "distributed solar" and utility-scale systems is one of both size and commercial arrangement. The electricity produced by utility-scale systems is typically sold to wholesale utility buyers, not end-use consumers. Technologically there are three main approaches to the large-scale conversion of solar energy to electricity at a large scale. The first and oldest is called concentrating solar power (CSP) or more precisely solar thermal electric generation. This technology uses mirrors to focus sunlight on an absorber where the solar energy is converted into heat, which is used to generate electricity. The other two technologies for large-scale solar energy conversion involve photovoltaics, the direct conversion of light into electricity that is the subject of most of this book. Concentrating photovoltaics (CPV) use lenses, mirrors, or combinations of the two to focus sunlight onto small specially designed PV cells. These systems need to accurately track the movement of the sun, and since they only accept a narrow angle of light only work well in areas where there is a high proportion of direct rather than diffuse sunlight. The other PV approach, so-called "flat plate" uses PV designed to operate at normal "one sun" conditions which is the type used in most other PV applications and represents the vast majority both of the installed base of solar generation as well as new capacity now being installed.

In the early days of terrestrial applications in the 1970s, PV systems as small as tens of kilowatts were considered utility scale if grid connected. Today systems as small as 1 MW are considered utility scale in some contexts but more typically utility scale refers to systems of at least 5 MW in size. Although a number of systems in the >100 kW range and connected to the grid were built in the 1970s, they were typically set up to power a specific facility. Arguably, the first real central station utility system was the 1 MW dc plant built, owned, and operated by ARCO Solar on the property of Southern California Edison's Lugo substation in Hesperia, California, which began operation on December 15, 1982.[29] This system covered 20 acres and consisted of 108 computer-controlled dual-axis trackers, each containing 256 monocrystalline PV modules rated at 36 W each.

Shortly thereafter, ARCO Solar built the larger 6.5 MW system on 160 acres owned by Pacific Gas and Electric (PG&E) on the Carissa Plains east of San Luis Obispo, CA. This system included planar mirrors adjacent to the modules to provide a low concentration of sunlight and pseudo-square cells to provide higher packing density and thus higher module efficiency. The system consisted of 756 two-axis tracking pole mounted arrays each containing 5 kW of single crystal silicon PV with the side mirror enhancement and an additional 43 larger trackers without the mirror enhancement.[30]

While the two systems' economics were improved by the tax credits available to solar energy equipment in the early 1980s in the United States, their costs were not made public and Arco Solar did not claim the plants produced power competitively with conventional generation. Rather the plants demonstrated that central station PV could be built at a large scale in relatively short periods of time and that the plants' output matched reasonably well with the demand peaks of the two respective utilities.[31]

These plants were dismantled in the early 1990s but provided experience in the engineering and operation of large-scale PV installations, the interface with the grid, and long-term PV module performance and reliability that were valuable to the industry and inform current module and plant design.

In 1984, not too long after these plants were constructed, the Sacramento Municipal Utility District (SMUD) built a similar 1 MW array on the grounds of the Rancho Seco nuclear power station (Fig. 2.10). "PV1," as it was called, contained 28,672 modules configured into 896 panels mounted in a single-axis east-west

[29]JC Arnett, LA Schaffer, JP Rumberg, REL Tolbert (1984). *Photovoltaic Solar Energy Conference; Proceedings of the Fifth International Conference*, Athens, Greece, October 17–21, 1983 (A85-11301 02-44), D. Reidel Publishing Co., Dordrecht, p 314–320.

[30]T. Moore (December 1985). Photovoltaics: Pioneering the solid-state power plant, *EPRI J*, p 19.

[31]ibid.

tracking configuration. The site was later expanded to 3.2 MW and remained in operation for 27 years, providing valuable information and validation of long-term PV performance.[32]

Figure 2.10 The SMUD PV System at Rancho Seco, 1992 (courtesy of Warren Gretz/NREL).

In other parts of the world, utility-scale installations in the 1980s were not as large but were nonetheless important in proving the ability of PV to be deployed at scale and contribute significantly to the supply of electricity. Germany's 300 kW installation on the Pellworm Island in 1983 has undergone several expansions and is still in operation as a hybrid system with 771 kW of PV, a 300 kW wind turbine, a 560 kWh Li-ion battery, and a 1.6 MWh redox flow battery. Although the original PV array was replaced in 2004, the system remains an important proving ground for PV technology.[33]

Japan was also an early leader in the deployment of PV, including one of the largest early utility-scale plants. In 1986, The New Energy Development Organization (NEDO) funded a 1.1 MW system at Saijo City, which included 1.8 MWh of battery storage to provide a more stable flow of power to the grid.[34]

[32]http://www.dupont.com/content/dam/dupont/products-and-services/solar-photovoltaic-materials/solar-photovoltaic-materials-landing/documents/SMUD_Solar_Power_Plant%20_Case_Study.pdf.

[33]Tanja Peschel (August 27, 2015). Germany's first large-scale photovoltaic plant is being refurbished, *Sun & Wind Energy*.

[34]Thomas B Johansson, Henry Kelly, Amula K N Reddy, Robert H Williams (ed), Laurie Burnham (executive ed) (1993). *Renewable Energy: Sources for Fuels and Electricity*, Island Press, p 500.

Deployments of utility-scale PV from the late 1980s through the 1990s were typically modest in comparison to the large demonstration projects of SMUD and Arco Solar at the beginning of the 1980s, with the exception of ENEL's 3.3 MW Serre (Italy) plant constructed in 1996. The objectives of these projects included comparison of various PV technologies, evaluation of inverters, understanding the dynamics of solar resource fluctuations and the resulting power output variation on the grid, and evaluating the potential cost savings in the distribution system.

In the 1990s, forward-thinking governments, most notably in Japan and Germany, began to create incentives for the deployment of grid-connected PV, typically starting with distributed applications, but the incentive mechanisms developed: capital cost rebates (via cash payments or tax credits), requirements on generating entities to produce a portion of their power from distributed sources, and feed-in-tariffs were adapted in the 2000s to energize the creation of the next wave of utility-scale PV plants.

2.5.1 Current Status

The three different technologies for large-scale solar generation have enjoyed different degrees of success since the beginning of the modern era of the deployment of solar energy in the 1970s. Their current status is discussed below.

2.5.1.1 Concentrating solar power systems

As noted in the introduction, CSP technologies use mirrors to reflect and concentrate sunlight onto an absorber where it is converted to heat. A working fluid carries this heat into a relatively conventional thermal power plant where the heat produces steam that then drives one or more turbines, which then turn generators to produce electricity. These systems are generally used for utility-scale projects. As of 2015, 118 operational CSP plants are deployed in 21 countries. The two main types of CSP systems in use today are parabolic trough and solar towers. The installed base of CSP plants is dominated by the parabolic through technology (around 80% of cumulative installed capacity), but now increasing numbers of solar towers are being built promising lower electricity cost. The capacity of the deployed systems ranges from 0.25 to 370 MW. In the United States, 2014 was the largest year ever for CSP—more than 767 MW of new CSP generation came online.[35]

Figure 2.11 The Solar Electric Generating Station IV power plant in California consists of many parallel rows of parabolic trough collectors that track the sun. The cooling towers can be seen with the water plume rising into the air, and white water tanks are in the background. Image courtesy of Sandia National Laboratory.[36]

Figure 2.12 Enlarged view of the parabolic troughs. Parabolic trough CSP—California's Mojave Desert 354 MW.[37] Image courtesy of NREL.

---

[35]http://www.nrel.gov/csp/solarpaces/index.cfm.

[36]http://www.nrel.gov/csp/solarpaces/parabolic_trough.cfm.

[37]http://www.nrel.gov/docs/fy11osti/48895.pdf.

Figure 2.13   Heliostats with silvered polymer reflectors surround the Solar Two power tower in Daggett, California. Image courtesy of Sandia National Laboratories.

By the end of 2014, worldwide 5 GW of CSP have been installed,[38] but the technology has two disadvantages, which are limiting its applications:

- Like all approaches that concentrate sunlight, only the direct portion of the solar resource is used. This means that in contrast to flat-plate PV systems, which produce electricity at any light level and respond nearly linearly, CSP does not perform well in locations where more of the sunlight is diffuse.

- The economics of CSP at this time is competitive only at certain locations. This is described in detail in chapter 3.3.1.

Although the technology is losing share to PV, the ability to store energy in the form of heat, which is much less expensive than energy storage in chemical batteries, allows these plants to provide power during late afternoon and evening periods when demand is highest in many areas, a valuable attribute that is keeping the technology alive. As noted in a recent report from IRENA: "Concentrating solar power (CSP) plants are just beginning to be deployed at scale. Although current costs are therefore high, cost reduction potentials are good and the ability to incorporate

---

[38]http://www.irena.org/DocumentDownloads/Publications/IRENA_RE_Power_ Costs_2014_report.pdf.

low-cost thermal energy storage will make them more important as the share of variable renewables in total power generation rises."[39]

2.5.1.2   Concentrating photovoltaic systems

As noted in the introduction, (CPV) uses mirrors and/or lenses to focus sunlight onto solar cells. Because the light is highly concentrated (most systems operate at hundreds of times the level of normal sunlight) specially designed PV cells are used to optimize performance under those conditions. In a concentrated solar cell system the solar cells always have to be at the focal point of the mirror or lens and therefore the system has to continuously track the sun so the sun's rays are kept perpendicular to the mirror or the lens, which requires a tracking mechanism. The tracker is an electromechanical device, which is programmed to follow the sun keeping the sun's rays perpendicular to the system.

To use sunlight concentrated by reflecting surfaces (mirrors) or lenses on a solar cell to produce electricity was considered as soon as solar cells were used for terrestrial purposes. There were several reasons why CPV systems were considered to be used. One of the reasons was that mirrors or lenses were cheaper at the beginning of the solar cell age to direct light onto solar cells than having the same area of solar cells to produce the equivalent amount of electricity. Today, solar modules are inexpensive which has removed this economic advantage for CPV. Another reason was that it was also found that the solar cells' efficiency is increased, thereby producing more electricity, by irradiating the solar cells with a light intensity many times that of the Sun.

Typical CPV systems using 300 to 1,000 concentration ratio require very accurate two-axis tracking machinery. As of the end of 2015, 54 CPV systems were operational globally. A small number were installed between 2006 and 1010, with the majority being deployed during 2011–2013. Most of the systems have capacities of less than 3 MW and only four systems were rated over 10 MW. The two largest (58 and 80 MW) are located in China. The number of installations decreased in 2014 and 2015 when some of the manufacturers went out of business due to competitive pressure

---

[39]http://costing.irena.org/technology-costs/power-generation/concentrating-solar-power.aspx.

from flat-plate PV. By the end of 2015 cumulative global CPV installations were 360 MW.[40]

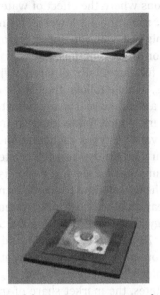

Figure 2.14 Using Fresnel lenses to concentrate the sunrays on a solar cell system.

2.5.1.3 Flat-plate photovoltaic systems: fixed and tracking

Photovoltaic (PV) systems either can have a fixed orientation or can incorporate mechanisms to track the sun. Fixed systems have the advantages of simplicity, lower cost, and packing density, while tracking systems offer greater energy generation per unit of capacity, and a broader curve of energy production over the course of a day. The economics of which is the best choice for a given project therefore depend on the relative costs of the PV and power-dependent components, the cost of the fixed and tracking mounting system options, the cost of the land, the value of electricity generated, and the characteristics of the solar resource.

Tracking the sun increases the energy output of PV to the greatest extent in areas where the proportion of solar energy

[40]Simon P Philipps, Andreas W Bett, Fraunhofer Institute for Solar Energy Systems ISE; Kelsey Horowitz, Sarah Kurtz, National Renewable Energy Laboratory (February 2016). *Current Status of Concentrator Photovoltaic (CPV) Technology—Version 1.2*, Freiburg, Germany & Golden, Colorado, USA.

coming directly from the sun (direct normal irradiance [DNI]) is highest in relation to the overall solar resource. Areas with high DNI are desert regions where the effect of water vapor and clouds in diffusing the solar radiation are lowest. This means that tracking systems have the highest value in these regions and tend to be chosen more often for projects in those areas.

There are two fundamentally different types of PV tracking systems: single-axis, which follow the sun's path from east to west typically with a fixed or seasonally adjusted elevation angle, and dual-axis trackers, which follow the sun in both altitude and azimuth. Although various types of actuation mechanisms have been developed, current utility-scale PV systems are all of the "active" type, meaning that an electric motor is used to drive the system typically with control input from one or more sensors.

In high-DNI areas, dual-axis tracking systems can produce up to 40% more energy than fixed-tilt systems. Single-axis systems can capture up to 75% of this increase and so represent the optimal configuration in a majority of situations today where trackers are economically attractive.

In the United States, the market share of tracking PV systems has been increasing and in 2014 represented 58% of new utility-scale systems and 41% of new utility-scale capacity (MW). The vast majority of these use single-axis tracking.[41]

Globally, fixed-tilt structures are still predominant accounting for 67% of utility-scale ground-mounted installations according to IHS Global Research.[42] This is due to the prevalence of fixed-tilt systems using seasonal adjustment in areas with low-cost labor, but single-axis tracking systems are expected to gain share in these areas as existing local suppliers develop products and new suppliers are expected to enter the market.[43]

In 2016, the largest PV systems are over 500 MW in size[44] and a rapidly growing market exists for large-scale solar PV

---

[41]Mark Bolinger, Joachim Seel, Energy Analysis & Environmental Impacts Division, Lawrence Berkeley National Laboratory, *Utility-Scale Solar 2014: An Empirical Analysis of Project Cost, Performance, and Pricing Trends in the United States*, LBNL-1000917, September 2015.
[42]http://press.ihs.com/press-release/technology/global-single-axis-tracker-revenues-expected-reach-nearly-2-billion-2019-ih.
[43]ibid.
[44]Denis Lenardic, http://www.pvresources.com/en/pvpowerplants/top50pv.php.

generation with over 22 GW installed in 2015, up from 14 GW in 2014.[45] This is created by a combination of policy support in the form of both economic incentives and a supportive regulatory environment for central station PV, where the markets exist, and fundamental economics, which are now at parity or below most other forms of generation in favorable conditions (good sun, low cost of capital, long-term commitments to purchase power).

Figure 2.15 Large PV Arrays as Seen From Space. On the left is the Desert Sunlight solar project in California's Mojave Desert, which became operational on February 15, 2015. The 550 MW plant covers an area of 16 km^2, and contains 8.8 million cadmium telluride photovoltaic modules. On the right is the similarly rated Topaz Solar Farm in San Luis Obispo County, CA, that became operational in June, 2014. Each image covers an area of 10.5 × 12 km. Credit: NASA/METI/ AIST/Japan Space Systems, and US/Japan ASTER Science Team

A list of the world's largest solar plants can be found at the PV Resources Web site (see footnote 41). Large US projects can be found on the Web site of the Solar Energy Industries Association (SEIA).[46]

The cumulative installed capacity of large-scale PV generation by country as reported by WikiSolar at the end of 2015 is listed in Table 2.2.

---

[45]http://wiki-solar.org/library/public/160307_Utility-solar_2015_figures_top_ 60GW.pdf.
[46]http://www.seia.org/research-resources/major-solar-projects-list.

Table 2.2 Cumulative utility-scale PV installations at the end of 2015 by country

Country	Plants	MWAC
China	522	18,975.6
United States	679	12,958.6
United Kingdom	508	4,520.6
India	330	4,183.3
Germany	298	3,613.2
Japan	98	1,804.2
France	136	1,629.5
Spain	172	1,524.7
Canada	119	1,508.8
South Africa	29	1,222.5
Italy	100	1,004.8
Thailand	81	1,000.5
Chile	17	859.8
Ukraine	16	499.7

Source: Wiki-Solar.

Asia, led by China, now dominates in the capacity rankings, a result of continued rapid growth in large-scale PV deployment driven by demand for electricity, desire to lessen the environmental impacts of generation, and very low costs for PV hardware and installation.

PV costs have now reached levels nearly unimaginable only 10 years ago. In many parts the world the installed cost of utility-scale PV is now well below \$1/W—the goal established by the US DOE's Sunshot program for the year 2020. When combined with superior solar resources, low cost of capital, and long-term power purchase commitments, the prices bid for solar PV generation have recently dropped below \$0.03/kWh—the lowest of any source.[47]

2.5.2 Future Prospects

The impact of the low electricity prices achieved by PV systems is revolutionizing the 100-year-old electric utility business

---

[47]http://www.bloomberg.com/news/articles/2016-05-03/solar-developers-undercut-coal-with-another-record-set-in-dubai.

(Chapter 5.3.) but returning to the question of whether central station PV can in fact "power the world," the International Energy Agency (IEA) has performed an extensive analysis of the potential of Earth's deserts for PV generation and has found that with conservative assumptions of 15% efficiency, 50% coverage area coverage, and 0.7 performance ratio, the deserts of the world could generate $2,239 \times 10^3$ TWh (=8,060 EJ), or 14 times of the world primary energy demand 560 EJ in 2012. Put another way using only 8% of the land area (with 50% coverage ratio) of the world's deserts could provide all of mankind's current energy needs. Even with growth of human energy demands in the coming century, solar PV has the potential to be a major part of that supply. A truly solar world powered by both distributed and central station PV plants—once seen as an idealistic vision—is now seen as likely, not only by advocates but by the businessmen and investors who are building this new world of solar energy (see Chapter 6).

2.6 Important Large Market: Solar Energy and Clean Water

Allan R. Hoffman

2.6.1 Desalination and Disinfection: Introduction

Water and energy are as basic as it gets—access to both is critical to poverty reduction, sustainable economic development, and national security. They are also inextricably linked—energy is needed to ensure delivery of water services and water is needed to produce fuels and energy. Water is essential to life—unlike the various forms of energy there are no substitutes—and as we progress into the 21st century global demands for both water and energy are increasing. A conflict arises because water and energy policies often conflict—burning fossil fuels to create more fresh water, as is traditionally done, releases carbon dioxide and other pollutants into the atmosphere, with adverse health, economic and environmental consequences. This creates an opportunity for use of solar and other forms of renewable energy to replace fossil fuels in this critical and increasingly important role as global demand for fresh water continues to grow. This chapter discusses two ways in which solar energy can be used to resolve this conflict.

2.6.2 Desalination

Solar energy has long been used for desalination (desalinization), the removal of dissolved salts from salty water to produce drinkable (potable) water. References to desalination using sun-heated water that evaporates and is then condensed, leaving the salt behind, can be found in historical records going back centuries. This process of evaporation and condensation is called distillation. Variations are widely used at sea to this day and helped keep many early explorers and traders alive during long ocean trips.

Solar energy's connection to water, aside from sun-heated distillation, is that solar photovoltaic (PV) modules can be used as an electricity source to pump water from underground aquifers in remote locations lacking grid connections (see chapter 5.1), and both flat PV modules and concentrating solar power (CSP) can be used as an energy source in place of fossil fuels, the usual energy source used today in large-scale desalination facilities. Steadily decreasing costs of and increasing experience with solar have made this possible. An irony of past practices is that in our attempts to provide clean water by burning fossil fuels to provide the needed heat and electricity to drive desalination processes, we are releasing carbon dioxide and creating climate conditions that adversely affect rainfall patterns, our primary source of clean water.

However, one might ask: Why is desalination important? Isn't Earth a water-rich planet to the tune of about 300 million cubic miles of water, with each cubic mile containing more than one trillion gallons? The problem is that most of that water, approximately 97%, is in the oceans, which have an average salt concentration (salinity) of 35,000 parts per million by weight, and drinking that water regularly can kill us. Ingestion of salt signals our cells to release water to dilute the salt. Too much salt, and this process will deplete your cells of moisture, your kidneys will shut down and your brain will become damaged. Eventually you will die without access to fresh water.

What about the water that is not in the oceans? Three percent of 300 million cubic miles is still a lot of water. Unfortunately, most of that 3% is not easily available for our use. Some is tied up in icecaps and glaciers, some is tied up as water vapor in the atmosphere, and the rest is in groundwater, lakes, and rivers. The other reality is that some of our freshwater supply is simply inaccessible due to its location and depth. The net result is that we make productive use of less than 1% of our global water resources.

We need desalination because the world's population, economic development, and the related demands for fresh water are growing. Agriculture accounts for three quarters of global water use. Fresh water is not distributed uniformly around the globe and we are contaminating existing sources of fresh water with farm and industrial runoff. In addition, farmers are over-pumping underground sources of fresh water in China, India, and the United States. What is also true is that several areas in Africa—

e.g., the Sahara, the Sahel, West Africa, Namibia, and Libya—have large fossil water reserves underlying their deserts that are still untapped. Libya is building the world's largest water pipeline from its desert to exploit this resource and deliver water to its capital and other coastal cities. And in California, a recent study identified deep underground water resources that were previously unknown, potentially tripling groundwater volumes in California's Central Valley, which is experiencing a serious drought.

Current annual global demand for fresh water, which has more than tripled in the past 50 years, is estimated to be about 1,000 cubic miles (1,100 billion gallons), approximately 30% of the world's total accessible fresh water supply. Under business as usual it has been estimated that this fraction could increase to 70% by 2025.

Too little fresh water has health, economic, and environmental impacts. The World Health Organization estimates that more than one billion people lack access to clean water supplies, more than two billion lack access to basic sanitation, and water-borne diseases account for roughly 80% of infections in the developing world. There are also gender implications of fresh water shortages. Women head one-third of the world's families, are responsible for half of the world's food production, and produce between 60% and 80% of the food in most developing countries. To produce adequate sanitation and food, they must first "produce" water. As the principal providers of water, women and girls in developing countries spend up to 8 hours daily finding, collecting, storing, and purifying water. This reduces significantly the time they might otherwise use for education, involvement in their communities, and income-producing cottage industries.

Where fresh water supplies are inadequate, desalination becomes a necessity. Significant advances in desalination technology started in the 1900s and took a major step during World War II because of the need to supply potable water to military troops operating in remote, arid areas. By the 1980s, desalination technology was commercially viable and commonplace by the 1990s. Today there are more than 16,000 desalination plants worldwide, producing more than 20 billion gallons of drinkable water every day. This is expected to reach more than 30 billion gallons per day by 2020, with one third of that capacity in the Middle East.

There are quite a few technologies today for removing salt from saline water in addition to direct sun-heated distillation. The most widely used are reverse osmosis (RO), multistage flash distillation (MSF), and multi-effect distillation (MED). MSF and MED are sophisticated versions of distillation where heat for evaporation is provided by burning fossil fuels. For example, Saudi Arabia uses approximately 300,000 barrels of oil every day to desalinate seawater providing 60% of its fresh water supply. Qatar depends on desalination of seawater to produce all of its water, deriving heat from the combustion of natural gas.

Reverse osmosis (RO), the most common type of desalination in use today (65% of installed capacity), requires no thermal energy, just mechanical pressure (800–900 psi) to force salty water through a membrane that separates the salt from the water. Electricity is needed to create this pressure, creating an opportunity for solar PV to provide the needed electricity. Energy-wise, RO is the most efficient of today's desalination technologies, requiring 3.0–5.5 kWh per cubic meter of fresh water produced, depending on the salinity of the source, seawater or brackish water. Saline water comes in different strengths and is categorized as follows: highly saline water: 10,000–35,000 ppm; brackish water: 1,000–10,000 ppm; fresh water: less than 1,000 ppm. Brackish water can be found in underground fossil water aquifers and is created when seawater invades fresh water supplies. Unfortunately, this is occurring more frequently today as sea levels rise due to global warming.

While most attention today is focused on large-scale solar-powered RO facilities, quite a few small-scale facilities do exist. One example is the project in Jordan jointly funded by the US DOE and the US Agency for International Development (AID), with in-kind contributions from Jordan, Israel, and the Palestinian Authority (PA). Initial discussions took place in 1997 and implementation occurred in two phases due to the unsettled situation between Israel and the PA. In Phase 1, a refurbished US Army ROWPU (Reverse Osmosis Water Purification Unit) was provided to Jordan for placement in the remote village of Qatar, located 300 km south of Amman, 35 km north of Aqaba, and close to the Dead Sea. The village, selected by the Jordanian Ministry of Water and Irrigation, had 35 houses, 250 inhabitants, and occasionally hosted troupes of nomadic Bedouins. It had no fresh water supplies—all its water was trucked in once a week from Aqaba and stored in a concrete block cistern. What it did have was an underground brackish water aquifer (3,865 ppm) easily accessible at a depth of 50 meters.

The ROWPU unit, powered by a diesel generator, was installed and maintained by the National Energy Research Center of Jordan. In addition to gaining experience and training operators, the purpose of using the ROWPU was to study the technical characteristics of the RO system to help in the design of a new system that would be powered solely by PV. An additional goal was to facilitate regional cooperation among Jordan, Israel, and the PA in the design and eventual local manufacture of small desalination units suitable for use in the many remote, water-stressed communities throughout Jordan and in the West Bank.

The final design pumped the brackish water from the well to a storage tank and passed it through the desalination unit with two exiting branches, one as potable water that was stored in a special tank and another as brine water (elevated salinity) that was directed to an evaporative pond. The produced water exited as fresh water permeates at a salinity of 30 ppm. An external mixing process at the exit produced water for consumption at a salinity level of 300 ppm.

The PV system was designed to power the unit to operate from 8:00 am to 4:00 pm and produce 16 gallons per minute of fresh water. Simulation assumptions and results were as follows: average daily solar radiation: 6.1 kWh per square meter per day; 140 PV modules with a peak rating of 16.8 kW; electrical load of the RO unit: 9.2 kW; average daily PV produced electricity: 95.4 kWh; 24 storage batteries with a total energy capacity of 73 kWh; average daily feed brackish water: 37 cubic meters per day; produced fresh water: 22 cubic meters per day.

This unit was to be installed on King Abdullah's palace grounds in Aqaba as a permanent demonstration for the Jordanian people, but eventually was installed at an industrial site just outside the palace grounds due to bureaucratic resistance from an official of the US State Department. Another unit was planned for placement in a PA village on the West bank, and parts were ordered and stored by the Israelis, who also trained the PA operators. However, the unit was never installed due to the intervention of the second Israeli-PA Intifada.

Large-scale solar-powered desalination was discussed extensively at the 2013 World Future Energy Summit in Abu Dhabi, UAE (United Arab Emirates). The future arrived in 2015 when Saudi Arabia's newly formed company Advanced Water Technology (AWT) announced that a 15 MW solar array will be used to supply 60,000 cubic meters of desalinated seawater a day to the city of Al Khafj in north-east Saudi Arabia at a cost of $130 million. The desalination unit is part of the Mohammed bin Rashid Al Maktoum Solar Park, scheduled to become the largest solar power plant in the Middle East in 2030 at 1 GW. The 13 MW Phase I of the project started operations in 2013. The 200 MW Phase II is scheduled to get under way in 2017.

In 2016 a French company, Mascara, announced that it will use off-grid rooftop solar to power a desalination project in Ghantoot, Abu Dhabi, that will be funded and operated by the clean energy group Masdar. It will be the fifth desalination unit located in Ghantoot, alongside four other grid-connected units that will run until 2017, with the aim of finding out "how the desalination plant of the future will operate" (Dr. Alexander Ritschell, Masdar Senior Manager). It is anticipated that most new small- and large-scale RO units will be solar powered where solar radiation is abundant as solar PV costs continue to drop.

In addition, a new desalination technology using the heat generated by a 400 kW concentrated solar power system (parabolic troughs) has been developed and tested and will be used in California's Central Valley to desalinate and reuse agricultural runoff. It can also desalinate other types of contaminated water. The technology used is similar to that of a parabolic trough power plant except that the heat generated will be used to drive a concentrated solar still that will evaporate and distill water at 30 times the efficiency of natural evaporation. Another advantage is that the system concentrates the removed salts into a solid so that no liquid brine residue becomes a disposal problem, a common issue for traditional desalination plants. The solid residue can be mined for useful mineral byproducts. A pilot facility already exists, and a full-scale facility is being designed to help address California's serious drought situation.

2.6.3 Disinfection

Solar water disinfection uses solar energy to make water contaminated by bacteria, viruses, protozoa, and worms safe to drink. It does this through some mix of PV-generated electricity, solar thermal heat, and solar-powered ultraviolet (UV) radiation. One common procedure is to create ozone, O_3, an unstable form of oxygen, by an electrical discharge in air or pure oxygen (O_2). The discharge dissociates the oxygen molecules into oxygen atoms, which combine with other O_2 molecules to create the ozone, which leads to oxidation and destruction of the pathogens. Generally used in medium to large disinfection facilities, it is more effective than chlorination but also more costly. Widely used in Europe, it is the least used disinfection method in the United States.

Direct exposure to UV radiation can also kill pathogens by disrupting their DNA and RNA, preventing their reproduction. This concept was used in 1993 by Dr. Ashok Gadgil of Lawrence Berkeley National Laboratory (LBNL), who invented UV Waterworks as a means of disinfecting contaminated water using solar-powered UV radiation. He was motivated by an outbreak of cholera in India, his native country, and focused on developing a technology that would be inexpensive and easily maintained without a skilled operator. It works by passing unpressurized water under a UV lamp, which does not come in contact with the water. The lamp can be powered by a single PV module or another source of electricity. It has long been known that UV radiation in the wavelength range 240–280 nm has this herbicidal effect and recent research seems to pinpoint 260 nm as the most biologically active wavelength. Special UV lamps used in this application put most of their energy into this wavelength region. A standard UVWaterworks unit can disinfect about one ton of water per hour at a cost of about five US cents. An exclusive license for manufacture and sales has been granted by LBNL to International Health, Inc. and units are now in use all over the world.

One further comment on use of solar-powered UV radiation lamps for disinfection: solid-state light-emitting diode (LED) technology has now been extended to the UV wavelength region and would be more energy efficient and potentially more reliable than the broader spectrum UV lamps that have been used so far. If UV LEDs can be developed for narrow band emission centered at 260 nm (efforts are under way) and can be produced inexpensively, they should be attractive replacements for UV lamps in future UVWaterworks or similar disinfection units.

2.6.4 Conclusion

As solar energy steadily becomes less expensive and more widely used, and as world population and demand for fresh water increases, solar-powered desalination and disinfection will become increasingly important parts of water supply systems in the 21st century.

To estimate the potential market for PV associated with these applications, let us assume that each installed kilowatt peak of solar PV operates for 8 hours per day at an average of 0.7 kW. This produces 5.6 kWh of electrical energy per day. If half of the new fresh water capacity anticipated from desalination in the next four years (as stated above, an increase from 20 billion to 30 billion gallons per day) is PV-powered at 4 kWh per cubic meter of fresh water (a cubic meter equals 264 gallons), this will require (5 billion gallons/day) × (1 cubic meter/264 gallons) × (5.6 kWh/cubic meter) = 7.58×10^7 kWh of PV-generated electricity. Under the above assumptions, this corresponds to 13.5 million installed peak kilowatts, which at an installed cost of $2/peak watt ($2,000/peak KW) is a total cost of $27 billion. If only a quarter of the new desalination capacity is PV powered, this still amounts to $13.5 billion, or $2.7 billion for every billion gallons per day of increased capacity. Even if installed PV costs drop by a factor of two in future years, this amounts to a respectable PV market, especially as more and more desalination is put in place and is PV-powered.

One final estimate of market potential: the UVWaterworks unit requires 1 kWh per day to deliver clean drinking water for 2,000 people and more than a million units are now in operation. At one million units, this requires 1 million kWh per day, and at 5.6 kWh per installed kilowatt peak per day and $2,000 per installed kilowatt peak, this corresponds to an invested cost of $357 million, a reasonable PV market as well.

2.7 Quality and Reliability of PV Systems

John Wohlgemuth[48]

PV has no moving parts. It just stands motionless producing electricity and therefore requires practically no maintenance. It is also well accepted today that PV modules are warranted for 25 years to produce a predicted amount of electricity. This extremely important feature, that it will produce the predicted amount of electricity for a quarter of a century, is not even questioned by the detractors of the PV systems, the electric utilities, lobbyists of the fossil fuel, and nuclear industries in spite of the fact that are no or very few man-made machines that carry a quarter of a century guarantee. This long-term guarantee is extremely important because it is one of the reasons why PV electricity is cheaper than the electricity produced by nuclear energy or by coal.

As mentioned, the long-term reliability as well as the safety of PV products and systems was a factor in achieving the commercial success of the electricity produced from solar energy. Investors would not risk billions of dollars to purchase and install PV systems if they did not think that these systems would produce enough electricity for sale to make them a profit. This chapter will review the history of how PV gained its reputation as a reliable source of electricity and then discuss what is being done today to further improve long-term performance.

Much of this chapter will talk about the PV modules themselves although the modules are usually the most reliable part of the PV system. However, modules are also the most expensive part and by far the hardest and most expensive to replace if something goes wrong. Today most PV modules are warranted for at least 25 years with a maximum allowable degradation rate of 0.8%/year. Of course just having a warranty may not be of much value if the company goes out of business. Where do you go today to make a warranty claim for a failed company like "Evergreen Solar" module? While a warranty is a commercial decision made by the company, one would hope that there is a technical basis for that decision. In most cases, module types do not have 25 years of field experience. Therefore, we have to rely on what limited

---

[48]John Wohlgemuth's biography is on page 293.

field data is available as well as the results of accelerated stress testing to predict module survival (qualification testing). Even if a module type is capable of survival, actual survival depends on the module manufacturer's Quality Management System (QMS) and the quality of its installation.

Other components in the PV system do not have the same level of reliability history. Usually around 50% of all reported system failures are due to the inverters.[49] For inverters you need quick support when it fails. Therefore, it is important to purchase your inverters from a reliable source with good warranty and customer service support. If the inverter goes down it needs to be fixed or replaced as soon as possible to minimize the lost revenue.

2.7.1 Module Qualification Testing

The issue of the necessity to make sure that solar cells and modules have long lifespan goes back to 1958 when it became obvious that the entire space satellite program depends on the life of its power source: the solar cells and solar modules. Research was undertaken in the United States, Europe, and Japan and solar cell and module qualification tests were developed and it was determined that a manufacturer's QMS was needed.

In 1975, the US government initiated a terrestrial PV research and development project, one of the aims of which was to help the terrestrial PV industry to produce reliable PV modules. The management of this project was assigned to the Jet Propulsion Laboratory (JPL) in Pasadena, California. At JPL, John V. Goldsmith, who was for years involved in the quality assurance program for the space solar cells and modules, became the manager of this project. He wanted to adapt the space solar cell and module quality requirements to the totally different terrestrial conditions, establishing a specification to achieve this goal. A quality and testing program[50] was established to buy "blocks" of terrestrial PV modules in meaningful quantities from manufacturers but

---

[49]Bradley Hibbard, NREL PVMRW 2011 (http://www.nrel.gov/docs/fy14osti/60170. pdf).

[50]A detailed description of the entire JPL "Flat-Plate Solar Array Project – Final Report" by RG Ross and MI Smokler is available: JPL Publication 86-31 (http://authors.library.caltech.edu/15040/1/JPLPub86-31volVI.pdf).

their products must qualify in each subsequent block to more and more stringent requirements. The JPL Block I–V program was very successful as it had a big effect on the quality and reliability of the PV modules.

The subsequent JPL sets of accelerated stress tests exposed many weaknesses in module construction. They taught the early PV industry that to survive in the field, modules had to be able to pass at least this minimum set of accelerated stresses. To pass these tests, the manufacturers had to radically redesign their products. Comparing the modules from Blocks I and II with those from Blocks IV and V, one can see that the designs changed dramatically in this short time period.[51] Many of the designs adopted in Block V are still in use in today's PV modules. As pointed out by Peter Varadi,[52] JPL also required module manufacturers to utilize a QMS.[53] The module designs from Block V became engrained in the culture of some of the top-tier module manufacturers of those days.

So how successful were these early qualification tests? One study by Whipple[54] looked at module performance during their first 10 years of operation. Whipple reported that for the module types not qualified to Block V, 45% failed in their first 10 years of operation. For module types qualified to Block V, less than 0.1% failed in their first 10 years of operation. Ask yourself where the PV business would be today if we still had 45% failure rates in the first 10 years of operation.

Based on the Block V specification, the International Electrotechnical Commission (IEC) published IEC 61215: Crystalline silicon terrestrial photovoltaic (PV) modules—Design qualification and type approval in 1993. This was followed in 1996 by publication of IEC 61646: Thin film terrestrial photovoltaic (PV) modules—Design qualification and type approval. These rapidly replaced the JPL Block V and other national and regional standards around the world. The third edition of IEC 61215 was

---

[51]Melvin I Smokler, David H. Otth, Ronald G Ross Jr (1985). The block approach to photovoltaic module development, *18th IEEE PVSC.*

[52]Peter F Varadi (2014). *Sun above the Horizon*, Pan Stanford Publishing, Singapore.

[53]The requirements of the JPL introduced Quality Management System was very similar to what the International Organization for Standardization (ISO) introduced only in 1987, ten years later.

[54]Marjorie Whipple (1993). The performance of PV systems, *NREL/DOE PV Performance and Reliability Workshop*, Golden, CO.

published in March 2016. It combines the qualification tests for crystalline silicon and thin films into one series of standards with some new requirements.

Since the early 1980s, many of the customers have been requiring that the PV modules they purchase have passed a qualification test. Any module you purchase should be qualified to either IEC 61215 or IEC 61646.

2.7.2 Module Safety Certification

It is clearly important that modules perform safely over the course of their lifetime. The first module safety standard was established by the American Underwriters Laboratories (UL), UL 1703: "Standard for Flat-Plate Photovoltaic Modules and Panels," first published in 1986. It addresses electrical and fire safety and uses accelerated stress tests similar to those in JPL Block V to evaluate whether the modules remain safe during their years of operation. UL1703 has been written into the US National electric Code (NEC) and is therefore required by law in most places within the United States.

The IEC addressed module safety after the publication of the UL document. Safety is in general more of a national (or regional) issue than qualification or performance, so trying to develop a consensus safety document that meets the needs of most nations around the world can be difficult. For example, in the United States per the NEC all electrical systems are grounded. In Europe most electrical components are double insulated and not grounded. Trying to write a component safety standard that meets both needs can be very difficult. After many years of trying, the IEC finally published IEC 61730: Photovoltaic module safety qualification in 2004.

The goal had always been to harmonize UL 1703 and IEC 61730 so that module manufacturers do not have to qualify to both, but because of National differences in safety requirements the effort to harmonize UL 1703 with the first edition of IEC 61730 was abandoned. A second edition of IEC 61730 that matches better the requirements of the US National Electric Code should be published in late 2016. An effort is now under way to harmonize UL1703 with this second edition of IEC 61730.

The original version of UL 1703 contained fire tests that basically tested and rated PV modules as though they were roofing material. In 2014 UL completely changed the way modules are fire rated. It is really now a test of the module mounting system. If the mounting system allows flames to get under the modules, it will fail. If it can deflect the flames up away from the roof, it can pass the fire test.

While most PV modules today are certified to either UL 1703 or IEC 61730 or both, it is important to recognize that these safety certifications tell you nothing about the reliability or durability of a PV module. There is no requirement in either document for the modules to perform adequately after the stress testing. Passing is based entirely on the stressed modules not presenting safety hazards before and after the testing. A module producing no electricity is typically safe but not very useful in your PV system. To state it simply UL 1703 and IEC 61730 are not replacements for IEC 61215. If you want a reliable module, it should be also qualified to IEC 61215.

2.7.3 Module Warranties

One interesting measure of how the module manufacturers feel about the reliability, durability, and lifetime of their products is to look at what warranty they are offering. Table 2.3 shows how the warranty period increased on Solarex power modules over the years. While this table is specifically for the Solarex warranty, most other crystalline silicon module manufactures offered similar warranties. In 1993, I presented a paper at an NREL/DOE Performance and Reliability Workshop entitled "Testing for Module Warranties."[55] In the paper, I explained the accelerated tests that Solarex performed before extending the warranty from 10 to 20 years.

The year before Solarex increased its warranty from 10 to 20 years, its major competitor Siemens Solar increased their warranty from 10 to 15 years. This helps to point out that warranties are ultimately commercial decisions. Solarex didn't just match Siemens's warranty but offered a better warranty than

---

[55]John H Wohlgemuth J (1993). Testing for module warranties, *NREL/DOE PV Performance and Reliability Workshop*, Golden, CO.

Siemens. In this case, the Solarex technical staff was able to provide technical justification for the longer warranty. That is the critical issue with a warranty; the manufacturer should be able to provide data from the field and from accelerated stress testing that validates the claim that their modules can survive at least as long as the warranty period.

Table 2.3   Solarex warranty periods for crystalline silicon modules

Date	Length of warranty
Before 1987	5 Years
1987 to 1993	10 Years
1993 to 1999	20 Years
Since 1999	25 Years

What do the warranties promise? Historically there were three parts to a PV module warranty.

1. A workmanship component saying that the module would continue to look nice during the specified time period, originally only a year, but now typically 10 years or more.
2. An intermediate power guarantee: With the 25-year warranty, this usually warrantied no more than 10% power loss over 12 years.
3. A final power loss guarantee, typically no more than 20% power loss over the length of the warranty.

This system worked fine as long as most of the modules were being supplied to standalone or small-size grid-connect systems where meeting the load was the main goal. When people began to build PV systems to sell the electricity, they did not like the step function warranty. For their energy calculations, they assumed that the first 10% drop occurred immediately and that the drop to 20% occurred in the 12th year. This sort of energy calculation meant the PV modules were not worth as much as the module manufacturers thought. To counter this, the larger module manufacturers began offering yearly warranty levels—guaranteeing no more than 0.8% power loss per year for 25 years. This is the sort of warranty you should get if you are going to buy a large number of modules.

2.7.4 Failure Rates in PV Systems

Almost every study on PV reliability shows that PV modules are one or the most reliable components in PV systems (as can be seen on Fig. 2.16, it is only 3%). Balance of system (BOS) components, as indicated previously, especially inverters, have the highest failure rates. Often approximately half of all reported service events related to a PV system are caused by the inverters. Figure 2.16 shows the cumulative data from several studies of PV system maintenance requirements.[56] So why are we worried about the modules? It is all a matter of degree. When an inverter goes down, it does shut down the output power, but often it can be fixed by a simple reset or replacement of a component or board. On the other hand, when a module fails, it shuts off the whole string and usually must be replaced by a spare. If more than just a small fraction of the modules fail, it is likely that the whole PV system will have to be replaced, taking the system down for a long time and even if covered by warranty still likely to cost a lot to replace.

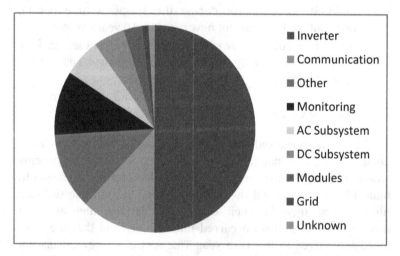

Figure 2.16 Cause of service cases for PV installations.

---

[56]Bradley Hibbard (2011). *NREL PVMRW* (http://www.nrel.gov/docs/fy14osti/60170. pdf); Tassos Golnas (2011). *Sandia National Labs Utility Scale Workshop* (http:// energy.sandia.gov/wp-content//gallery/uploads/Golnas-SunEdison_Operator.pdf).

I published papers on failure rates for Solarex and BP Solar modules using warranty return data.[57,58] From 1994 to 2005, the total number of module returns was 0.13% of the Solarex and BP Solar modules in the field. In the later years (2006–2008) the annual rate of return for BP Solar multicrystalline silicon product dropped to approximately 0.01%.

2.7.5 Module Durability Data

Modules not only fail, but they also degrade over time. D. C. Jordan[59] has studied the reported module degradation rates in the literature. Figure 2.17 comes from one of his papers. The median degradation rate is 0.5% per year with an average of 0.8%. One would assume that systems with significant numbers of failed modules are likely not included in this data, so we do not see any degradation rates above 4%. Based on the literature, a majority of these PV modules are meeting their power degradation warranty of less than 0.8% per year.

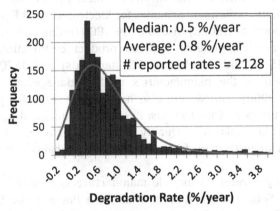

Figure 2.17 Degradation rates for PV modules reported in the literature.[11]

[57]John H Wohlgemuth, Daniel W Cunningham, Andy M Nguyen, Jay Miller (2005). Long term reliability of PV modules, *20th EU PVSEC*.

[58]John H Wohlgemuth , Daniel W Cunningham, Dinesh Amin, Jay Shaner, Zhiyong Xia, Jay Miller (2008). Using accelerated tests and field data to predict module reliability and lifetime, *23rd EU PVSEC*.

[59]Dirk C Jordan, Sarah R Kurtz (January 2013). Photovoltaic degradation rates—an analytical review, *Prog PV* **21**, 12–29.

2.7.6 ISO 9000

You cannot test quality into a product. Testing can only indicate that the manufacturer was successful in building quality into the product. JPL not only instituted accelerated stress tests of the modules but also required the manufacturer to have a QMS. In the early days of PV, each module manufacturer set up their own QMS. However, by the mid-1990s, most PV module manufacturers had their QMS certified under ISO 9000. So not only did you want a module that was qualified to IEC 61215, but you wanted it produced in an ISO 9000–certified factory.

2.7.7 IECQ and IECEE

In 1996, an updated version of the JPL Block V system for PV modules and components was established under International Electrotechnical Commission Quality Assessment System for Electronic Components (IECQ).[60] This system was subsequently transferred to International Electrotechnical Commission System of Conformity Assessment Schemes for Electrotechnical Equipment and Components (IECEE) in 2005. The IECQ/IECEE system for PV modules and components included product certification under IEC 61215 (or other relevant IEC standards) and IECQ/IECEE certification of the manufacture's quality management system (e.g., ISO 9000). The system also had requirements for periodic product retests and inspections. PV products qualified under the system could display a PV Quality Mark that customers should be able to distinguish them from untested and probably less reliable products.

Initially several PV module manufacturers, including Solarex TATA BP Solar (renamed recently as Tata Power Solar Systems Ltd.), ASE Applied Solar Energy GmbH and Websol (India), obtained IECQ/IECEE certification on their products. The system was abandoned because many governments did not specify that only "quality" products could be used for government-supported projects. This is especially true in Germany, where the government did not specify that for customer protection only "quality" product would qualify to participate in the FiT program in spite of the fact that in 2004 about 25% of the modules offered were untested.

---

[60]Peter F. Varadi (2014). *Sun above the Horizon*, Pan Stanford Publishing, Singapore.

Only the World Bank supported the IECQ/IECEE quality system and recommended that in the programs they supported, IECQ/IEEC approved PV products should be used.

In China, the "Golden Sun Mark,"[61] the quality requirement of which is similar to IECQ/IEEC system, was adopted by the "China General Certification Center" (CGC) for PV products in 2006. Hundreds of PV products have been qualified under this program.[62]

2.7.8 To Further Improve Long-Term Performance:

IEC 61215 Qualification testing and ISO 9000 Factory Certification—Why isn't this system good enough today?

For a number of years PV modules qualified to IEC 61215, built in an ISO 9000 factory were the norm. This system defined quality products. However, when customers started building really large (multi-tens of megawatts) PV systems they began wondering if indeed this was enough. After some analysis it became clear that indeed it was not enough.

- Looking at Figure 2.17, we see that there are a significant number of modules in the tail to the right of the figure. A number of modules have degraded more than 0.8% per year and so would not meet the 25-year warranty.
- In the 1990s, most of the major module manufacturers (Solarex, Siemens, and then Shell Solar, Sharp, Kyocera, etc.) used additional accelerated stress testing beyond the levels prescribed in the qualification tests to ensure product quality and to validate that those products could meet the warranty, but many newer module manufacturers did not.
- PV systems have changed significantly since the 1990s. Transformerless inverters resulted in ungrounded systems that can lead to potential-induced degradation (PID), a failure mode unknown 10 years ago. Higher PV system voltages are themselves causing new module problems.
- PV cells and modules are always changing. Thinner but larger cells are more prone to breakage. Larger modules and

---

[61]http://www.cqc.com.cn/www/english/servicebydepartments/new-energy-certification-department/golden-sun-certification/.
[62]Peter F. Varadi (2014). *Sun above the Horizon*, Pan Stanford Publishing, Singapore.

the components within them suffer from higher mechanical stress levels.

- There is a continuing effort to squeeze as much cost as possible out of the module. Sometimes manufacturers cut corners that they should not have cut, resulting in premature module failures.

Many PV system installers and purchasers began to develop their own set of extended accelerated stress tests. Usually the additional tests were based on increasing the stress levels of the tests in IEC 61215. The philosophy seemed to be that if a 1,000-hour damp heat test was a good, then 2,000 or even 3,000 hours would be better. The problem with this logic is that often the failure modes observed after these extended tests were not failure modes observed in the field. If PV modules are to operate in a damp heat chamber, then an extended damp heat test would be a good one. However, modules are usually designed to operate outside in the sunlight. The number one rule of selecting a valid accelerated stress test is that it must duplicate failures seen in the field. Many of these extended test sequences did not meet this criterion. For example, the failure mode observed after 3,000 hours in typical glass superstrate/EVA/polymer backsheet crystalline silicon modules is the corrosion of the conductive oxide that provides the ohmic contact between the silicon and the silver grid lines. This corrosion is caused by a build-up of humidity within the module to levels that can never be reached in the terrestrial environment. So this failure mode has not been observed for fielded modules. So why are some customers requiring module designs that can pass this test? It will only add unnecessary cost to the product.

In 2012, at the NREL PV Module Reliability Workshop,[63] a session was held in which different PV groups were invited to present their extended accelerated stress test regime. In this session entitled "IEC 61215 on Steroids,"[64] more than 10 different organizations presented their extended test protocols. Of course, no two of the test protocols were the same and none of the organizations were able to justify their tests via demonstration that their tests identified field failure modes that were not identified by the qualification tests themselves.

[63]Technical Report NREL/TP-5200-60169, November 2013.
[64]NREL PVMRW 2012 (http://www.nrel.gov/docs/fy14osti/60169.pdf).

At the same time that these groups were developing their own extended test protocols, some large purchasers of modules began requiring module manufacturers to have their factories inspected by specialists hired by the customer. Since there was no standard that covered such inspections, companies like Solar Buyer developed their own set of criteria.

So for each project, a module manufacturer may have to have their modules tested to the customer's extended test protocol and their factory inspected by the customer's agent. This, of course, adds appreciably to the cost and likely adds little value since there is really no technical justification for many of the added test requirements. The factory inspections may add value if they help the manufacturer improve their quality management system. However, there is no standard for these inspections; so there is no guarantee that they will improve anything and it is very likely that they will all be different.

2.7.9   International PV Quality Assurance Task Force

To help address the issue of how to best test for PV module wear-out, Sarah Kurtz and John Wohlgemuth from the NREL, United States; Masaaki Yamamichi and Michio Kondo, Advanced Industrial Science and Technology (AIST), Japan; and Tony Sample, Joint Research Center (JRC), Europe, with the assistance of James Amano, Semiconductor Industry (SEMI), organized the International PV Module Quality Assurance Forum. The forum was held in July 2011 in San Francisco with more than 150 PV reliability experts from around the world participating. This event fostered international participation in a new organization created at that forum, entitled the International PV Quality Assurance Task Force, called PVQAT[65] for short (pronounced as "PV CAT"). The reliability of PV modules was the first issue this Task Force was to tackle.

At the forum, two goals were set for the organization:

1. To gain a better understanding of how PV modules fail in the field and to develop a set of accelerated stress tests that would duplicate those failures in the laboratory.
2. To provide guidelines to PV module manufacturers to assist them in developing improved quality management systems.

[65]http://www.pvqat.org.

A third goal related to development of a conformity assessment system has since been added. This will be discussed in detail later.

PVQAT is open to participation by anyone who wishes to contribute. The effort relies on the research done by volunteers around the world. PVQAT helps to guide worldwide research to answer important questions related to testing that predict outdoor performance of PV modules. The objectives of this work are to develop IEC standards that can be used for testing PV modules that will be deployed in any terrestrial environment. PVQAT established three goals:

- PVQAT Goal #1—Improved Reliability testing
- PVQAT Goal #2—Quality Management System for PV Module Manufacturing
- PVQAT Goal #3—IECRE (IEC Renewable Energy)

PVQAT Goal #1—Improved Reliability testing

The PVQAT teams are looking for field failures of module types that have successfully passed the qualification tests. So this is an effort to identify failures caused by module wear-out. The initial guidance in this effort came from a paper by Dirk Jordan of the NREL.[66]

Based on these observations, PVQAT selected the following degradation/failure modes to develop new or improved accelerated stress tests for

- Module discoloration

  Encapsulant discoloration is the most observed field degradation mode and was responsible for much of the power loss in early EVA[67] type modules, particularly those from Arco and Siemens. Improvements in the EVA formulations have led to much more stable encapsulants that do not discolor. To ensure that future encapsulants do not discolor, an IEC standard will be issued shortly.

- Diode failures

  Various modes of diode failure have been identified. These failures can be eliminated or reduced by improving

---

[66]Dirk C. Jordan, John H. Wohlgemuth, Sarah Kurtz (2012). *27th EU PVSEC.*
[67]EVA (ethylene vinyl acetate) is used in the *photovoltaics* industry as an encapsulation material for *solar cells* in the manufacture of *photovoltaic modules.*

the specification, instituting a training program for the installations of the diode as part of the QMS and adding one new diode test to the IEC qualification tests.

- Solder bond and interconnect ribbon failures due to thermal cycling

Recently Nick Bosco from the NREL studied the stress on solder bonds caused by thermal cycling either in the field or in a test chamber.[68] He found that the stress was worst at high temperatures and when the local weather had the largest number of temperature reversals during the day due to e.g., passing clouds. So in some places like Sioux Falls, South Dakota, and most of northern Europe, the module never gets hot enough to cause extreme stress on the solder bonds. In those types of climates, the qualification test of 200 thermal cycles is probably adequate to test for a 25-year warranty. However, in hot climates with a significant number of passing clouds during the hot part of the day like in Phoenix, AZ or Chennai, India 200 thermal cycles is nowhere near enough to simulate 25 years of field exposure. In those climates 500 to 600 cycles are necessary. So the climate specific IEC standard will have to include more thermal cycles for those climates.

- Cracked cells

Cells can break in the module at almost any part of the life cycle from manufacture to shipping to installation to deployment. The use of electroluminescence (EL) as a tool in PV is a great method for observing cracked cells. Many module manufacturers EL scan every module before shipping. Large numbers of broken cells are caused by poor handling practices during installation of the modules. This can include everything from dropping, twisting, and walking on the modules. This is one of the reasons for wanting to start the IECRE Conformity Assessment system and to establish an IEC standard on proper PV system installation practices. Finally, the IEC is planning to add a dynamic

---

[68]Nick Bosco N, Timothy J Silverman, Sarah Kurtz (2016). Climate specific thermomechanical fatigue of flat plate photovoltaic module solder joints, *Microelectronics Reliability*, **62**, 124–129. (http://www.sciencedirect.com/science/article/pii/S0026271416300609).

mechanical loading test to the qualification tests to assess a module's susceptibility for broken cells.

- Potential-induced degradation

Potential-induced degradation (PID) refers to degradation in module performance that is due in part to the voltage of the PV system. The PID degradation can happen for several reasons. The IEC has published a technical specification (IEC 62804) that provides two methods of measuring the PID sensitivity of a PV module. At present, manufacturers can test their product to that standard, to be soon included in the IEC 61215 module testing standard.

- Delamination and corrosion

When delamination is accompanied by corrosion of the cell metallization or interconnect ribbons, it can lead to significant power loss. This failure mode has been recognized as one of the most important by PVQAT, but as of now it is still unclear what causes this problem. More research is needed to better understand this degradation process. Ultimately, a test that causes this type of delamination will have to be added to the climate specific test sequence.

- Failures due to higher-temperature operation

The IEC qualification tests have all been written for moderate temperatures. IEC 61215 says it is suitable for operation in general open air climates, which have a maximum ambient temperature of 45°C. The IEC safety standard is even more restrictive saying it is suitable for operation between –40 and +40°C. Of course, there are places on Earth, such as Saudi Arabia, where it routinely gets hotter than 40 or 45°C. Then, of course, there are rooftop arrays where module temperatures as high as 100°C have been measured. An IEC document entitled "Guideline for Qualifying PV Modules, Components and Materials for Operation at Higher Temperatures" is planned to provide one source for information related to improved testing of modules and other components such as junction boxes, connectors, and cables for operation at higher temperatures.

PVQAT Goal #2—Quality Management System for PV Module Manufacturing

The second goal of PVQAT was to provide guidelines to PV module manufacturers to assist them in developing improved quality management systems. To facilitate this effort, four groups from around the world (United States, Japan, China, and Europe) reviewed the ISO 9001 standard and made recommendations about how to improve it for PV module manufacturing. Once each group had finished their work, a worldwide team under the direction of Paul Norman and Ivan Sinicco began to review and combine the four lists into a coherent document. These were all incorporated into a document entitled "Proposal for a Guide to Quality Management Systems for PV Manufacturing: Supplemental Requirements to ISO 9001-2008," which was published as an NREL report.[69]

The NREL report was turned into a technical Specification for the IEC, which was published in January, 2016 as IEC TS 62941. So this Technical Specification is now available for manufacturers to use in certifying their quality management system.

PVQAT Goal #3—IECRE

The third goal of PVQAT was to develop a conformity assessment system for the PV systems deployed in the field. The IEC Conformity Assessment system has had the capability to cover PV modules through IECQ and then IECEE for a number of years as was discussed previously. In addition, IECEE has accredited PV test laboratories for many years. However, there has been nothing in the IEC conformity Assessment system to cover actual energy producing systems in the field. The IEC conformity Assessment System was set up to certify widgets not power plants.

The first to realize this deficiency were the member of the IEC Technical Committee on Wind Energy led by Sandy Butterfield (who would later become the first chairman of the IECRE Management Committee). Sandy wished to have standards written for installed wind turbines. However, the IEC said that certifying wind turbines was a conformity assessment responsibility, but

---

[69]Norman P, et al (2013). *Proposal for a Guide to Quality Management Systems for PV Manufacturing: Supplemental Requirements to ISO 9001-2008*, NREL Technical Report No. TP-5200-58940. http://www.nrel.gov/docs/fy15osti/63742.pdf.

there was no conformity assessment system available under the IEC to cover this. So they went ahead breaking the IEC rules and writing the Quality Management documents for wind energy systems anyway. It is often easier to ask forgiveness than to get permission beforehand. Once the documents were completed, they went back to the IEC for permission to set up a new Conformity Assessment System that could cover wind farms.

In the meantime, Sandy Butterfield contacted the representatives of IEC PV Technical Committee, including Howard Barikomo. Howard helped rally the PV community to join the wind group to petition the IEC to establish a new Conformity Assessment system that would cover renewable energy power plants. An important meeting was held in Oslo, Norway, in October 2012 during the IEC General Meeting. Among the representatives at the meeting from PV were Howard Barikomo, Heinz Ossenbrink, and I. Among the representatives of the IEC was Kerry McManama, who would eventually become the executive secretary of IECRE. This was an extremely important meeting as the IEC staff seemed to finally realize why the systems they had in place (IECQ, IECEE and IECX) were not well suited to extend conformity assessment to power plants, but that the new renewable energy sectors (wind, PV, and marine) needed such a system.

In 2014, the IEC approved the creation of IECRE (for renewable energy). It was created with three sectors, wind, PV, and marine energy. Additional renewable energy sectors such as solar heating would be welcome if they see a need to join.

The motivation for the IECRE system is to provide confidence that a PV plant will safely perform as promised while reducing the cost. The ICERE system is designed to streamline the due diligence process, leveraging what others have already learned. Investors want the highest quality product at the lowest possible price. We need balance in this equation and a system that will optimize both. To provide for flexibility in this equation IECRE has set up several levels for assessment. IECRE covers a variety of systems sizes with very different requirements based on size:

- U1: Utility Scale (> 1 MW)
- U2: Residential (< 25 kW)
- U3: Commercial (200 kW to 1 MW)
- U4: Collection of small systems

In addition, the IEC standards being written to cover the procedures and measurements required for the IECRE system allow for different levels of accuracy in the measurements. In this way a customer can pick the level of assessment and measurement accuracy they want for their particular system.

Balancing cost and perfection requires a system built for this. The IECRE PV sector led by Chairman Adrian Haring with major support from George Kelly and especially Sarah Kurtz has worked diligently to set up the goals and operations of the PV sector to achieve this. According to Sarah Kurtz,[70] the principles that ICERE has used to achieve this balance are (1) to seek standardization, (2) provide oversight at every stage of the project, (3) emphasize consistent quality control, and (4) provide efficient implementation, e.g., don't duplicate inspections. In order to stream line the IECRE process, it has been broken down into steps, each of which can be certified by the IECRE process. Four main steps are design qualification, substantial completion (system is ready to operate), annual performance, and asset transfer (the health of a PV plant as a basis for an acquisition).

The PV committee is now in the process of preparing the documents that describe the Rules of Procedure for IECRE, how the IECRE PV system will operate. These documents can be found on line at http://www.iecre.org/documents/refdocs/. These OD documents refer to the relevant standards that are to be used in certifying PV power systems. These include

- Module standards (IEC 61215, IEC 61730 and IEC 62941);
- PV plant design guidelines (IEC 62548 for small systems to be published in 2016 and IEC 62738 for utility-scale systems to be published in 2017);
- System installation standards (IEC 62446-1 on commissioning with a new edition recently published and IEC 63049 on Quality Management for the installation process to be published in 2017);
- PV system measurement standards (IEC 61724-7 a Capacity test to be published in 2016 and IEC 61724-3 an energy test recently published).

---

[70]Sarah Kurtz (July 2016). Defining bankability for each step of a PV project using IECRE, *Intersolar North America*.

As of the middle of 2016, the IEC PV system is now far enough along that those wishing to serve as IECRE Certification Bodies or IECRE Inspection Bodies can apply. It is anticipated that the applications for the first certifications can be made during the second half of 2016, and with any luck the first certificate will be issued before the end of 2016.

Balance of System (BOS) Components

At the NREL Module Reliability Workshop held in Golden, Colorado, in February 2016,[71] the first day covered modules, the second day had breakout sessions on modules and systems, and the third day covered PV systems and inverters. During the third day, it became very clear that the BOS, in general, and inverters, in particular, were years behind modules in terms of reliability testing and quality management in manufacturing. While there is a PV inverter qualification test standard, it was borrowed virtually test for test from the module qualification standard without considering how inverters fail in the field. It was seldom if ever mentioned during the inverter sessions at the Workshop.

One of the results of the workshop was the decision to expand PVQAT into the components of the balance of system, e.g., module-integrated electronics (MIE), connectors, etc.

The result of the US government's terrestrial PV quality and reliability program initiated in 1975—which even today is called the "JPL program"—was that the electrical output of properly manufactured PV modules can be guaranteed for 25 years. This program became one of the pillars of the success of the PV systems. The continuation of that work resulted in further improved quality and reliability. Present and future work not only for the PV modules but also for the entire PV system has paramount importance as the issue is now the easy and inexpensive financing—"bankability"—of PV systems in sizes, e.g., for millions of rooftops to huge utility-size systems.

---

[71]NREL PV Module Reliability Workshops (http://www.nrel.gov/pv/pvmrw.html).

2.8 Storage of Electrical Energy

Allan R. Hoffman

2.8.1 Introduction

Energy storage, the capture of energy produced at one moment for use at another moment is not a new concept. When mankind discovered fire, it also discovered that the heat and light emitted when wood was burned was somehow stored in that wood until magically released. The need for storage to steady the output from a variable energy source is also not new. In 1861, the following words appeared in an agrarian newsletter: "A Mighty Wind One of the great forces nature furnished to man without any expense, and in limitless abundance, is the power of the wind. Many efforts have been made to obtain a steady power from the wind by storing the surplus from when the wind is strong." Today, our understanding of energy and its storage is more advanced, new storage technologies are coming into play for mechanical, thermal, chemical, and electric energy, and storage is becoming an essential element of our evolving energy system. This chapter will review the basis for this emerging role for energy storage, the many forms it can take, the various benefits it provides, its economics and status today, and its future potential. It is a set of technologies critical to an energy system increasingly dependent on intermittent energy sources such as wind and solar energy. It can also provide important benefits to any energy system where storing energy at one time and releasing it at another has important system stability and economic implications. The chapter's primary focus will be on electricity storage.

2.8.2 Why Is Electrical Energy Storage Important?

The Energy Storage Association, a US national trade association for the energy storage industry, answers this question on its Web site as follows: "Energy storage fundamentally improves the way we generate, deliver, and consume electricity. Energy storage helps during emergencies like power outages from storms, equipment failures, accidents or even terrorist attacks. However, the game-changing nature of energy storage is its ability to balance power supply and demand instantaneously—within milliseconds—which makes power networks more resilient, efficient, and cleaner than ever before."

Looking into history, ever since the discovery of electricity generation using rotating coils of wire in magnetic fields by the British scientist Michael Faraday in 1820, people have sought ways to store that energy for use on demand. Without such storage, or use in some other way (e.g., to electrolyze water to create and store hydrogen, heat water, bricks or phase change materials that store heat, or refrigerate water to create ice) surplus electricity generation is lost. With modern societies increasingly dependent on energy services provided via use of electricity, the need for electricity storage technologies has become crucial—think of life without your portable phone or computer. This is especially true as more and more intermittent (variable) renewable electricity enters the grid, to avoid grid destabilization. This can occur because electric power supply systems must balance supply and demand, and because demand is highly variable and hard to control the balancing is routinely achieved by controlling the output of power plant generators. If these generators are intermittent solar and wind, and their grid contribution becomes significant, achieving the balance is that much more difficult, and a means of compensating for these variations is needed

There is also strong economic and social incentive for storing electricity in a localized, distributed manner. Today's 100-year-old centralized utility business model, in which large central power plants deliver electricity to customers via transmission and distribution lines, includes the imposition of peak demand charges that can account for a significant fraction of a business' or an individual's electricity bill. With the use of localized generation (e.g., PV panels on the roof of your home or business location), combined with storage at your site, these peak charges can be reduced if not eliminated, and independence from the utility, to some degree, can be achieved. This reality is taking place today in Germany and Australia (and coming to the United States) and threatening utility business models. For example, in Germany several German utilities have recognized the new realities and have gone into the solar-energy storage business. They now sell, lease, and maintain roof-mounted PV and battery storage systems. Storage of electricity also offers other benefits of particular value to utilities. These benefits will be discussed in a later subchapter of this chapter

It is also important to recognize that as solar and wind generation have moved into the energy mainstream, largely due to cost reductions, increased experience with the technologies, and pressure to reduce carbon emissions from power generation, the interest in storage has increased accordingly. New applications of solar plus storage are moving from the analysis and small demonstration phase to a stage where bankable projects are being proposed and financed. The benefits of this marriage of technologies are system benefits and the number of storage products on the market is growing rapidly. Together with smart grid technology that facilitates efficient control of grid demand and allows solar and wind energy to be moved from often-remote locations where it is generated, to locations of high demand, it is the technology critical to full incorporation of our clean energy resources into our energy system.

2.8.3 What Are the Various Forms of Electricity Storage?

This question has many answers. Storage technologies can be categorized in different ways, depending on their scientific/ engineering basis, their key performance characteristics, duration of storage, and stage of development. The utility of these technologies in real-world applications derives from these categories. The following list divides the many diverse technologies into six broad categories:

1. Traditional and Advanced Batteries: a range of electro-chemical storage solutions, including advanced chemistry batteries and capacitors
2. Flow batteries: batteries where the energy is stored directly in the electrolyte solution for longer cycle life, and quick response times
3. Flywheels: mechanical devices that harness rotational energy to deliver instantaneous electricity

4. Superconducting magnetic energy storage: energy is stored in persistent magnetic fields
5. Compressed air energy storage: using compressed air to create an energy reserve
6. Pumped storage/hydro-power: using stored water to create an energy reserve
7. Thermal: capturing heat and cold to create energy on demand

Traditional batteries are those that have been around for a long time, starting with lead-acid batteries that are still the dominant battery storage technology today. They are widely used in cars and trucks and elsewhere because of low cost, high power density, and high reliability. Disadvantages are low specific energy storage capacity, large size and weight, and the need for an acid electrolyte. Lead is also a toxic material when inhaled or ingested. Research to improve lead-acid batteries has been under way for more than a century, and considerable progress has been made— e.g., improved lead-acid batteries that require no maintenance and widespread recycling of used batteries to recover the lead. Further progress is anticipated.

Sodium sulfur batteries, which operate at high temperatures (300–350°C), use molten sulfur as the positive electrode and molten sodium as the negative electrode. They are separated by a solid ceramic barrier that serves as the electrolyte. It was developed in the 1960s by the Ford Motor Company and subsequently sold to the Japanese company NGK. It has now been widely demonstrated in Japan and more than 270 MW of peak shaving capacity have been installed. US utilities are beginning to explore the technology for peak shaving, backup power, firming intermittent wind power, and other applications.

Nickel-cadmium batteries have been in commercial production since 1910. They are a traditional battery type that, while not known for high energy density or low first cost, provides a simple-to-manage, long-lasting, and reliable electricity storage solution. For many years, in small battery form, they were a primary electricity source for mobile devices.

Most battery attention today is focused on lithium-ion (Li-ion) batteries where cost and safety are prime concerns. The first commercial Li-ion battery was developed in Japan and released to the market in 1991. Initial applications were in consumer markets, but today many companies are developing larger-format cells for use in energy storage applications. These include their use in hybrid and fully electric vehicles, residential and business storage of solar-generated electricity, and multimegawatt containerized batteries for utility application.

Li-ion batteries are widely used today because "pound for pound they're some of the most energetic rechargeable batteries available." For example, it takes 6 kg (13.2 pounds) of a lead-acid battery to store the same energy as 1 kg (2.2 pounds) of a Li-ion battery. They also hold their charge well (today's Li-ion batteries lose about 5% per month), have no memory effect (removing the need to fully discharge before recharging), can handle hundreds to thousands of "round trips" (i.e., charge-recharge cycles), and have good round trip electrical efficiency.

The story does have a negative side—Li-ion batteries are sensitive to heat, cannot be fully discharged (thus requiring a computerized battery management system), and are still costly (although costs are coming down rapidly), and certain chemical formulations can occasionally burst into flame if damaged or otherwise overstressed. One person making a big bet on Li-ion batteries is Elon Musk, who has announced plans for a $5 billion battery factory, to provide Li-ion batteries for his Tesla electric vehicles and other applications. Through such large-scale production, Musk hopes to reduce the cost of the batteries by 30% to about $10,000 for a 60 kWh battery pack (Note: today's electric vehicles, depending on driving habits, get about 3–3.5 miles per kWh). Other Li-ion battery manufacturers are jumping into the market as well, e.g., Panasonic, which was the largest electric vehicle battery manufacturer in 2015, driving prices down even further, and many new applications using these batteries are being designed and deployed.

The term Li-ion refers not to a single chemistry but to a number of chemical combinations where lithium ions are transferred between the electrodes during the charge-discharge cycles. The lithium ions are derived from electrode materials that contain lithium compounds, and different compounds present different cell voltages, energy densities, life, and safety characteristics. Battery management systems are required—Li-ion batteries lack the ability to dissipate overcharge energy—and safety characteristics are a function of system design and control algorithms, regardless of battery cell chemistry.

Supercapacitors, another relatively new battery technology, which look like ordinary electrical capacitors but are usually bundled in packages, store energy in electric fields created by stored electric charge and fill a gap between ordinary capacitors and rechargeable batteries. Because the charge is stored physically, with no chemical or phase change occurring, the charge-discharge processes are fast and highly reversible. They can be repeated over and over again with virtually no limit at high efficiency. Depending on the design, supercapacitors (sometimes known as ultracapacitors) can have reasonably high energy densities of 6–20 Wh/kg. Because of their characteristics they are now widely used as low current power sources for computer memories and medical devices and in cars, buses, trains, cranes, and elevators, including energy recovery from braking. As a result, their market and applications are growing and the number of manufacturers is growing as well.

Flow batteries are large-scale rechargeable energy storage systems where rechargeability is provided by chemical compounds dissolved in liquids, which, when mixed together, generate electricity. A major advantage of flow batteries is that they can be recharged quickly by replacing the electrolyte liquid while allowing recovery of the active chemical components. They differ from conventional batteries in that energy is stored in conventional batteries as electrode material but as the electrolyte in flow batteries.

Redox (reduction/oxidation) flow batteries are particularly well suited to storing large amounts of energy—e.g., the surplus energy created by hours of solar or wind power generation—and are on the verge of wide application in the electric utility industry. The energy storage materials are liquids that are stored in separate tanks, and when energy is needed, the liquids are pumped through a "stack" where they interact to generate electricity. Many different chemical liquids have been tested for flow battery operation, with most current attention being focused on vanadium compounds, which are expensive. Flow batteries also have relatively low round-trip efficiencies and long response times. Because of the vanadium cost concern, many other chemical possibilities are being evaluated, e.g., zinc-bromine, zinc-chlorine, and iron-chromium flow batteries.

An important flexibility in the design of flow batteries is that the storage capability, i.e., the size of the storage tanks, can be tailored to the energy storage need of the particular application. The tradeoff is that the ratio of power to energy is fixed for integrated flow battery systems at the time of design and manufacture. They are well suited for a broad range of applications, with power requirements ranging from tens of kilowatts to tens of megawatts, and energy storage requirements ranging from several hundred kilowatt-hours to hundreds of megawatt-hours. They are also easy to control, flow can be stopped quickly during a fault condition, and can be easily scaled up.

Flywheels store energy by using electrical power to accelerate a cylindrical assembly called a rotor (flywheel) to a very high speed and maintaining the energy in the system as rotational energy. The energy is converted back to electricity by slowing down the flywheel. The flywheel system itself is a kinetic, or mechanical, battery spinning at very high speeds to store energy that is instantly available when needed.

At the core of most modern day flywheels is a carbon-fiber composite rim, supported by a metal hub and shaft and with a motor/generator mounted on the shaft. Together the rim, hub, shaft, and motor/generator assembly form the rotor. When charging (i.e., absorbing energy), the flywheel's motor acts like a load and draws power from the grid to accelerate the rotor to a higher speed. When discharging, the motor is switched into generator mode, and the inertial energy of the rotor drives the generator, which, in turn, creates electricity that is then injected back into the grid. Multiple flywheels may be connected together to provide various megawatt-level power capacities.

To illustrate the industry's capabilities, one major flywheel manufacturer offers a high-performance rotor assembly that is sealed in a vacuum chamber and spins between 8,000 and 16,000 rpm. At 16,000 rpm the flywheel can store and deliver 25 kWh of extractable energy. At 16,000 rpm, the surface speed of the rim would be approximately Mach 2—or about 1500 mph—if it were operated in normal atmosphere. At that speed, the rim must be enclosed in a high vacuum to reduce friction and energy losses. To reduce losses even further, the rotor is levitated with a combination of permanent magnets and an electromagnetic bearing.

An obvious issue associated with flywheels is catastrophic failure. With rotors moving at high rotational speeds and the flywheel structure experiencing large physical stresses, what would happen if a flywheel flew apart? The industry's answer is that they're designed for safety, which is probably correct, but most people still require additional reassurance. One solution would be to install the flywheels in sturdy containers or under building structures to minimize the impacts of a failure.

Advantages of a flywheel are high energy density and substantial durability that allows them to be cycled frequently with no degradation in performance. They also have very fast response and charge/discharge rates, being able to go from full discharge to full charge in a few seconds. They are particularly well suited for high power, relatively low-energy applications.

In a small size, flywheels are used by the US National Aeronautics and Space Administration (NASA) to point satellite instruments in the correct directions without the use of thrusters. In larger sizes, Beacon Power has several utility-scale flywheel storage projects operating in the United States: a 20 MW/5 MWh facility in Stephentown, New York (2011, 200 flywheels) and a similar 20 MW/5 MWh facility in Hazle, Pennsylvania (2014). A Canadian 2 MW flywheel storage facility by NRStor (10 flywheels) was put into operation in Minto, Ontario in 2014.

Superconducting magnetic energy storage (SMES) devices store energy in the magnetic field of a circulating dc electrical current in a superconducting coil. The superconductor has no electrical resistance and the current continues indefinitely unless its energy is tapped by discharging the coil. A typical SMES device has two parts, a cryogenic cooler that cools the superconducting wire below its transition temperature at which it loses its electrical resistance, and power conditioning circuitry that allows for charging and discharging of the coil. Its advantages are ultra-fast charge and discharge times, no moving parts, nearly unlimited cycling capability, and an energy recovery rate close to 100%. Disadvantages are cost of the wire, the need for continuous cooling, large area coils needed for appreciable energy storage, and the possibility of a sudden, large energy release if the wire's superconducting state is lost. SMES devices are often used to provide grid stability in distribution systems and for power quality at manufacturing plants requiring ultra-clean power (e.g., microchip production lines). One MWh SMES units are now common and a 20 MWh engineering test model is being evaluated.

Compressed air energy storage (CAES) utilizes surplus electricity to compress air to high pressures in underground caverns or other large storage vessels, which can then be heated and released as needed to power expansion turbines that generate electricity. One feature of compressed air storage is that the air heats up strongly (900 K) while being compressed from atmospheric pressure, 14.7 pounds per square inch, to storage pressures of about 1,000 pounds per square inch. Some of this heat can be removed by cooling to protect the multistage compressors or thermally stored before entering the cavern and used for subsequent adiabatic expansion of the stored air. Energy is also added to the compressed air during the expansion/ power generation cycle by heating with natural gas. Gases other than air, e.g., carbon dioxide, can be used as well.

In Europe and Argentina, city-wide compressed air energy systems were first built in the 1870s. The first utility-scale compressed air energy storage project was the 1978 290 MW Hunters plant in Germany using an excavated salt dome as the storage container. In 1991 a 110 MW plant with a capacity of 26 hours was built in McIntosh, Alabama. The world's third CAES project, opened in 2012, was a 2 MW facility in Gaines, Texas. More recently, the Utah-based Intermountain Power Project has announced a 1.2 GW CAES project in underground salt domes, with the first 300 MW to serve as storage for PV solar power and the next 900 MW as storage for anticipated new wind energy generation. The US DOE is also supporting several proposed CAES projects in Kern County in California and Watkins Glen in New York.

Pumped storage uses surplus electricity, usually at night, to pump water from a lower reservoir to a higher one, and then this water is allowed to run down through turbines to generate electricity as needed. Because of the large scale possible in such schemes, pumped storage is the most common type of utility storage today based on megawatts installed—22,000 MW in the United States alone.

In many ways pumped storage is similar to CAES in that surplus electricity is used to store energy in a large reservoir. It should also be noted that the material raised to a higher level does not have to be water. Companies are today revisiting a concept first proposed in the mid-19th century whereby a windmill would be employed to raise a quantity of iron balls, and these balls would then be allowed to fall into buckets on one side of a wheel, causing the wheel to rotate and thus drive a machine. Modern versions of this concept substitute gravel for iron balls and the mechanical system drives a turbine and generates electricity.

Thermal storage allows storage of energy in the form of heat or cold for use at a different time. Power-generating examples include modern solar thermal power plants, which produce all of their energy during daylight hours. Surplus energy produced during these hours can be stored thermally in the form of hot oil or molten salt, and other higher-temperature storage schemes are being explored. Another approach is to use off-peak electricity to cool water or create ice, which can be used in a building's cooling system to lower electricity demand during the day. Both types of thermal storage are in use today, and in growing amounts.

2.8.4 Applications of Energy Storage and Their Value

In addition to the large utility-scale applications mentioned above, energy storage systems can deliver a broad range of benefits to both utility customers and the grid. For customers these include backup power, increased self-consumption of PV-generated electricity, reduction of demand charges, and optimized management of time-of-use utility rates. For utilities energy storage provides a range of important support services such as frequency and voltage control, peak shaving, deferral of investments in distribution and transmission infrastructure, relief of transmission congestion, adequacy of supply, energy arbitrage, spinning/nonspinning reserve, and energy for black start after a shutdown.

Initial attempts at evaluating the value of storage have focused on single application examples and concluded that storage is too expensive today for widespread use. The World Energy Council's recent study, "E-storage—Shifting from Cost to Value," calls this conclusion into question. In its attempt to analyze "what the cost base of an array of storage technologies really means," it concluded that "a narrow focus on levelized costs of electricity

(LCoE) alone," which is the standard commonly used in measuring electricity generation costs, "can be misleading." The report shows that the LCoE metric when applied to energy storage fails to take into account the full range of values and revenue benefits offered by storage, and that the value offered by storage varies depending on the application.

This report followed up another recent study by the financial advisory firm Lazard, "Levelized Cost of Storage Analysis" (November 2015), which noted that battery economics are usually evaluated on the basis of single-use cases. This ignores the revenue benefits of multistacking, i.e., using batteries for more than one purpose, each with its own revenue potential. While evaluating these economic benefits gets difficult quickly, it can enhance battery economics and can be a game-changer for determining the financial viability of energy storage projects. The broad conclusion is that storage should be evaluated as a totally new element in the energy scene.

As the Lazard study points out, evaluating the economics of battery energy storage is difficult. Batteries are not strictly a supply- or demand-side solution but rather an arbiter of supply and demand, serving as either load or generation depending on whether they are charging or discharging. We know that batteries today are used for a small fraction of their useful lifetimes and can do much more if applications are stacked. Calculating the economics for stacked cases is high on Lazard's and other organizations' analytic agendas.

2.8.5 Capital Costs of Energy Storage

This is a rapidly moving target as more and more companies announce storage products and consumers and utilities begin to appreciate the full value of storage technologies. While the cost of energy storage has been coming down for 20 years, a great deal of new interest was generated on April 30, 2015, when Elon Musk announced his Tesla Powerwall rechargeable Li-ion battery product for home use. He advertised it at $350/kWh for a 10 kWh battery pack and $429/kWh for a 7 kWh version. He also offered a Powerpack, a utility-scale battery storage system at $250/kWh. These costs were considerably below anticipated costs before Musk's announcement. Other companies soon jumped into the market, announcing their own storage products, and the race was on.

Today, costs are falling and markets are expanding rapidly. It is anticipated that storage will be more disruptive to the energy system than solar, and Citigroup has cited $230/kWh as the price point "where battery storage wins out over conventional generation and puts the fossil fuel incumbents into terminal decline." UBS and Navigant, in a joint report, anticipate that the $230/kWh mark will be reached in broad markets within the next few years and storage costs will likely fall to $100/kWh. They also predict that, as a result, the market for battery storage will grow 50-fold over its 2015 level, mostly in helping households and businesses self-consume their solar generation, but also at grid-scale and through their use in electric and hybrid-electric vehicles This prediction is supported by the market research firm IHS which expects the energy storage market to "explode" to an annual installation rate of 6 GW in 2017 and over 40 GW by 2022, from an initial base of 0.13 GW installed in 2012 and 2013.

2.8.6 Concluding Remarks

There are many energy storage technology options that work and tradeoffs are often required—e.g., among storage capacity, power capacity, round-trip efficiency, and most importantly cost. However, "cost" has to be evaluated in terms of energy storage's many sources of value and many potential revenue streams. Lots of research is currently under way to develop new applications and reduce costs, given the large potential markets and the need to safely integrate variable renewable energy generation from solar and wind into the utility grid system.

What is becoming increasingly clear is that storage represents a fundamental change in our electrical energy system. Over the past 150 years, we developed grids throughout the world that immediately consume what they produce, and manage that by overproducing a little bit to make sure that backup exists in case of unforeseen outages, However, if you have energy storage, there is no need to overproduce and no need for backup reserves. It allows you to store electricity and use it when it is needed. This is the new world that energy storage is making possible.

2.9 Solar Energy and Jobs

Allan R. Hoffman

2.9.1 Introduction

As documented in other chapters in this book, solar energy is rapidly entering the energy mainstream. Many people commenting on this phenomenon attribute it to the role that solar energy can play in reducing carbon emissions from fossil fuel-powered electricity generation, and the fact that the costs of solar PV modules have come down dramatically in recent years. What has often been overlooked to date, and is just beginning to attract increased attention, is the fact that investment in solar and other renewable energy technologies can support economic development and job creation, more than traditional energy industries. This chapter discusses the evidence for this important conclusion.

2.9.2 What Are the Facts?

The first question to be asked in examining the impact of renewable energy development on jobs is: What data are available? A number of reports over the years have provided national figures for direct jobs in the various renewable energy industries, as well as the indirect jobs, e.g., in supply chains, that support these industries. What has been lacking until recently has been a regularly updated compilation of such direct and indirect jobs on a global scale. IRENA, which is discussed in a separate chapter of this book, has now filled this need with its Annual Reviews entitled "Renewable Energy and Jobs." These reviews have become the global standard for reporting job information related to renewable energy industries.

In the United States, such job information is available from several sources. Beginning in 2010, the US Bureau of Labor Statistics (BLS) received funding to start collecting data on so-called "green jobs." They established the BLS Green Jobs Initiative with three goals: "to develop information on (1) the number of and trend over time in green jobs, (2) the industrial, occupational, and geographic distribution of the jobs, and (3) the wages of the workers in these jobs." However, the BLS quickly learned by reviewing the literature that no widely accepted definition exists for this category of employment and so they created their own. According to the BLS, green jobs are either: "Jobs in businesses that produce goods or provide services that benefit the environment or conserve natural resources; or jobs in which workers' duties involve making their establishment's production processes more environmentally friendly or use fewer natural resources." The common thread linking these two parts of the definition "...is that green jobs are jobs related to preserving or restoring the environment." Included in this definition are jobs associated with producing energy using renewable energy resources.

Annual reports from the BLS provided data on green jobs for years 2009, 2010, and 2011. Unfortunately, this effort came to a mandated end on March 1, 2013, when federal budget sequestration went into effect and President Obama ordered across-the-board spending cuts required by the Balanced Budget and Emergency Deficit Control Act, as amended. Under the order, BLS had to cut its budget by more than $30 million and eliminate all "measuring green jobs" products.

Nevertheless, other organizations, including IRENA, have attempted to fill this gap. They include the Solar Foundation's annual "National Solar Jobs Census," monthly job information from the US Energy Information Administration, and occasional reports from the Environmental and Energy Study Institute, the Union of Concerned Scientists, the Global Green Growth Institute, NextGen Climate America, the United Nations Industrial Development Organization, and the International Energy Agency! Principal sources of information are periodic surveys of renewable energy industry employers. Collating this information has been difficult in the past because of differences in questions asked and job categories identified, but more recent efforts, especially by IRENA, is producing more reliable results.

What are these surveys telling us? The latest report from IRENA, "Renewable Energy and Jobs: Annual Review 2016," offers the following Key Facts:

- Global direct and indirect renewable energy employment in 2015, excluding large hydropower, reached 8.1 million, a 5% increase over 2014.

- Direct large hydropower jobs decreased slightly to 1.3 million.
- The increase in renewable energy jobs contrasted with job decreases in the broader energy sector.
- Most renewable energy jobs were in China (3.5 million), followed by Brazil, the United States, India, Japan, and Germany.
- Most job growth took place in Asia, which accounted for 60% of renewable energy employment.
- Solar PV, with 2.8 million jobs, was the largest renewable energy employer, an 11% increase over 2014.
- Solar PV jobs grew in Japan and the United States, stayed constant in China, and went down in the European Union.
- Jobs in solar water heating and cooling decreased to just under 1 million.
- Wind power had a record year in 2015, with employment reaching 1.1 million.
- Bioenergy in its various forms contributed 2.9 million jobs in 2015.
- Renewable energy employed more women, percentage-wise, than other parts of the energy sector.

According to the Solar Foundation the solar industry in the United States grew dramatically in 2015, growing by more than 20% for the third straight year and more than 120% since 2010. In fact, growth in solar sector employment in 2015 accounted for 1.2% of all new jobs in the United States. and significantly outpaced the overall national employment growth rate of 1.7%. Total solar jobs numbered just under 209,000, with 120,000 installation jobs (57.4%) topping the list. This number is expected to increase by 14.7% in 2016. Of the 2015 total, manufacturing accounted for 30,000 jobs (14.5%), sales and distribution 24,000 (11.7%), project development 22,000 (10.7%), and other solar jobs 12,000 (5.7%). More than 9,000 companies worked in solar energy, and more than 5 million US homes used some solar power.

California led the United States with nearly 76,000 solar jobs. It was followed by the other top 10 states as follows: Massachusetts: 15,095; Nevada: 8,764; New York: 8,250; New Jersey: 7,071; Texas:

7,030; Arizona: 6,922; Florida: 6,560; North Carolina: 5,950; Colorado: 4,998.

How do these numbers compare with employment numbers in the fossil fuel industries? These latter numbers are difficult to calculate because of the variety of direct and indirect jobs that are included in any survey of the field. Workers employed directly in the extraction of oil and gas in the United States at the end of 2015, who operate and develop oil and gas fields, numbered just over 187,000. Indirect jobs, which provide support for oil and gas operations such as exploration, excavation, well surveying, well and pipeline construction, account for just over two million additional jobs, 40% of which are reported to be in minimum wage jobs at gas stations. The number of jobs in several of these categories are also declining—for example, extraction lost nearly 14,000 jobs in 2015 and the pipeline construction industry lost 9,500 jobs. The coal mining industry, suffering from a significant reduction in coal-fired power generations in the United States in recent years, currently employs only 68,000 people.

It is important to recognize that fossil fuel technologies are typically mechanized and capital intensive, whereas the renewable energy industry is more labor intensive. Several studies have documented the fact that renewable energy creates more jobs per megawatt of power installed, per unit of energy generated, and per dollar of investment than the fossil fuel industry. For example, the 2010 Citizens Climate Lobby report "Building a Green Economy" by Joseph Robertson states: "The transition to a low-carbon economy creates the potential for a rapid and sustained expansion of jobs. Direct job creation for oil and natural gas is 0.8 jobs per $1 million in output, and coal's is 1.9 jobs per $1 million in output. Compare that to building retrofits for energy efficiency, which directly create 7 jobs per $1 million in output. Mass transit services create 11 and the smart grid creates 4.3. Wind, solar and biomass power generation, create 4.6, 5.4 and 7.4, jobs per $1 million in output respectively." An earlier study by the Renewable and Appropriate Energy Laboratory at Lawrence Berkeley National Laboratory (LBNL) produced similar results, as did a 2014 study ("Low Carbon Jobs: The Evidence for Net Job Creation from Policy Support for Energy Efficiency and Renewable Energy") by the United Kingdom's Energy Research Center.

Such numbers raise an interesting question, which appeared in the comments section of an article ("Over 3 Times More Green Jobs per $1 Invested Than Fossil Fuel or Nuclear Jobs") that appeared in *CleanTechnica* e-journal in 2013: "Energy is just a resource that allows us to accomplish real things. If green jobs require more workers isn't that a bad thing? It's like saying something breaks down more often and requires many more workers maintaining it." The easy response that was given is "If a clean energy source can produce electricity for about the same or less cost than fossil fuel/nuclear and employ more people, is that not a good thing? Seems to be a win-win and win some more."

What are the prospects for solar energy jobs in the future? Two recent reports address this issue. IRENA's report "Letting In The Light: How Solar PhotoVoltaics Will Revolutionize the Electricity System" states that "The age of solar energy has arrived. It came faster than anyone predicted and is ushering in a global shift in energy ownership. ...In only five years, global installed capacity has grown from 40 GW to 227 GW. By comparison, the entire generation capacity of Africa is 175 GW." Specifically, the report anticipates that solar PV could account for 8% to 13% of global electricity in 2030, compared to 1.2% at the end of 2015. Coming to a similar conclusion, Bloomberg New Energy Finance in a June 2016 report stated that "...solar and wind technologies will be the cheapest way to produce electricity in most parts of the world in the 2030s" and forecasts growth in solar PV reaching 15% of total electricity output by 2040. As IRENA Director General Adnan Amin said in a recent statement: "The renewable energy transition is well under way, with solar playing a key role." With respect to jobs in solar installation, Andrea Luecke, Executive Director and President of the Solar Foundation, has stated: "Growth for the next 10 years in solar is going to be very, very exciting." and as Lyndon Rive, CEO of SolarCity points out "You can't outsource these jobs."

The prospect of rapid growth in solar and other renewable energy technologies implies a corresponding drop in jobs in traditional energy industries. How are workers in these industries to fare in a world requiring different skills? As stated by Dr. Stephan Singer, director of Global Energy Policy for the World Wildlife Fund (WWF) "Workers benefitting from renewable energy expansion are often highly skilled, urban and flexible. That is very different from a coal miner or a gas pipeline worker. One does not make an offshore oil driller in the Gulf of Mexico into a solar PV engineer in San Francisco." Thus, as with any new or emerging technology, obsolete jobs will disappear and new jobs will be created. The critical need is to make the transition as painless and humane as possible, recognizing the human toll in lost jobs. This means investing in helping people develop new skills, which requires political will and financial support.

2.9.3 Concluding Remarks

If the US or other national goal is to create jobs in the energy sector, investing in clean energy, efficiency, and renewables is considerably more effective than investing in fossil fuel or nuclear technology. "Clean energy is no longer a niche business—it's a big time job creator. Our lawmakers need to realize that—and put policies in place, right now, to help the sector grow even more" (Dan Smolensk, managing director of The Green Suits, a talent recruitment and career development firm based in Virginia). It is clear that increasing use of solar energy is a powerful engine of economic growth and job creation. In the words of former Michigan Governor Jennifer Granholm: "Americans want good-paying jobs, and solar jobs are growing 12 times faster than the rest of the economy, our citizens are making and installing these solar panels, and with the right policies, the U.S. can create hundreds of thousands more solar jobs here at home. What more needs to be said?" The only thing is to not forget the people who will be displaced from their jobs in traditional energy industries.

3

Financing

3.1 Financing of PV

During the second half of the 1990s, Wolfgang Palz of the EU started programs and Hermann Scheer in Germany and Allan Hoffman of the US Department of Energy (DOE) organized conferences on "Financing of Renewable Energy." These programs and meetings resulted in two important findings:

- The widespread utilization of PV is mostly dependent on financing and much less on technology or cost.
- Private money would be available for a large number of and/or large-size PV systems. What would be needed is long-term security that the invested money and some profit would be assured.

As it was described in Chapter 2.2, the "Aachen model," which was enacted in Germany in 2000 as the Feed-in Tariff (TiF) law provided the long-term security and some profit for the money invested in PV systems. As it was expected, larger and larger amount of private money became available in Germany for PV Systems. As indicated in that chapter, it is estimated that in the following 10 years in Germany about $64 billion was invested in rooftop PV systems. This large amount of money apparently came from individuals' savings or from the loans taken by them or from investors to whom the property owner leased the roof space.

As a result, the demand for PV modules increased to an unbelievable level, which required factories to gear up for mass production and start new manufacturing facilities. The required huge amount of money these companies needed were from various stock markets and also from privates and banks. For example, 11 Chinese companies in the time frame 2005–2010 raised over $2 billion from the New York stock exchanges.[1] In addition, American and German companies also raised a substantial amount of money. The year 2005 was the time when the infamous Solyndra was able to raise a staggering $1.235 billion and went bankrupt in 2011.

[1]Peter F. Varadi (2014). *Sun Above the Horizon*, Pan Stanford Publishing, Singapore.

Sun towards High Noon: Solar Power Transforming Our Energy Future
Peter F. Varadi
Copyright © 2017 Peter F. Varadi
ISBN 978-981-4774-17-8 (Paperback), 978-1-315-19657-2 (eBook)
www.panstanford.com

The mass production of the many competing companies from 2010 to 2015 reduced the price of PV modules to well under $1.00/watt, which opened three incredibly large new markets: rooftop, commercial, and utility-scale PV systems It became evident that the widespread utilization of PV was mostly dependent on financing and much less on further development of technology or cost reduction.

Money for PV was available. What was needed was the development of the technology of financing geared to the unique requirements of PV. This needed government regulations and proper financing methods. This section of the book describes the results.

3.2 Subsidies and Solar Energy

Allan R. Hoffman

3.2.1 Introduction

In principle, subsidies are not difficult to understand—they are one of many policy instruments used by governments to achieve economic, social, and environmental objectives. In practice, they engender confusion because subsidies take many different forms and discussions of them can quickly get technical. A vast literature on subsidies exists—it is a popular topic among academic scholars and others. It just takes some effort to make sense of it all. This chapter will attempt to do so with respect to subsidies for energy technologies, and specifically for solar energy.

3.2.2 What Forms Do Energy Subsidies Take?

In their simplest forms, energy subsidies are financial measures that keep prices for consumers below market levels, to stimulate specific consumer behaviors, or keep prices for producers above market level to discourage certain business activities. Subsidies can also be used to reduce costs for both consumers and producers.

The US Energy Information Administration (EIA), formally a part of the US DOE, breaks subsidies down into five main categories:

- direct cash expenditures to energy producers and consumers
- tax expenditures via provisions in the US tax code
- R&D expenditures for increasing energy supplies or improving energy efficiency
- loans and loan guarantees for certain energy technologies
- electricity supply programs for specific geographical regions—e.g., the Tennessee Valley Authority (TVA)

Others include indirect support mechanisms such as tax exemptions and rebates, price controls, trade restrictions, and limits on market access. Taking all these categories into account, it is clear that today's modern energy industries have all relied on substantial subsidy support. That history in the United States is discussed below.

3.2.3 What Is the History of US Energy Subsidies?

Such subsidies are as old as the country itself. They date back to 1789 when the new government imposed a tariff on the sale of imported British coal. Over the intervening years, the coal industry has received many forms of assistance, usually federal and state tax breaks for mining and use, technological support for mining and exploration, national resource maps to encourage exploration and development, tariffs on foreign coal, and aid to industries that burn coal to encourage greater use and develop a steady market for coal.

A new phase in government support of energy technologies started in 1916 as the US love affair with the automobile began and Congress passed a new tax provision allowing oil companies to write off dry holes as well as all "intangible drilling costs" in their first year of exploration. Then in 1926 Congress approved the "depletion allowance," which lets oil producers deduct more than a quarter of their gross revenues. Both provisions were introduced as permanent parts of the US tax code. Production of liquid fuels from coal, to power our military activities, also received strong support during World War II (WWII).

Nuclear power, which grew out of the Manhattan Project's wartime research and development (R&D) of atomic weapons, was incentivized by the Atomic Energy Act of 1946. After the use of nuclear weapons to end WWII, there was a strong desire to find peaceful uses for the atom. The Act established the Atomic Energy Commission (AEC), which inherited all of the Manhattan Project's R&D activities and created a framework for government control of civilian nuclear power plants for electricity generation. Industry was concerned about potential liability in the event of a nuclear accident and the limited amounts of liability coverage offered by the insurance market. As a result, in 1957 Congress passed and President Eisenhower signed into law the Price–Anderson Act, which has been renewed several times since. It governs liability-related issues for all nonmilitary nuclear facilities constructed in the United States before 2026. The main purpose of the Act is to partially indemnify the nuclear industry against liability claims arising from nuclear accidents while still ensuring compensation coverage for the general public. The Act also helped firms secure capital with federal loan guarantees.

In the favorable environment created by these incentives, more than 100 nuclear power plants were built in the United States by 1973. In its latest renewal, the Act requires the nuclear industry to cover the first $12.6 billion of damages, with costs above that to be covered by retroactive increases in nuclear utility liability or the federal government. It is fair to say that civilian nuclear power industries would not exist in the United States or in many other countries without the Price–Anderson Act or its equivalent limiting utility liability.

For much of the early part of the 20th century, US federal energy incentives had been focused on increasing domestic production of oil and natural gas. In the post-WWII period, nuclear power entered the mix. As documented in a report by Fred Sissine, Specialist in Energy Policy at the Congressional Research Service (CRS)—"Renewable Energy R&D Funding History: A Comparison with Funding for Nuclear Energy, Fossil Energy, and Energy Efficiency R&D"[2]—total federal spending for fossil energy in the period 1948–1977 was about $16.4 billion in constant 2013 US dollars. Federal expenditures on nuclear energy, fission and fusion, amounted to $49.3 billion during that period.

The energy crises of the 1970s—the 1973–1974 OPEC Oil Embargo and the Iraq-Iran War—led to a new appreciation of energy issues and forced the US government to broaden its focus to include renewable energy and energy efficiency. Major changes included the establishment of the Federal Energy Administration (FEA) in 1974, and the Energy Research and Development Administration (ERDA) in 1975, which absorbed the responsibilities of the AEC and several energy programs in other federal agencies. The DOE was established on October 1, 1977, incorporating activities of the FEA and ERDA. All of the federal energy R&D programs—fossil, nuclear, renewables, energy efficiency—were brought under its purview, along with new programs in energy storage and electricity transmission and distribution (T&D) systems.

Between 1916 and 2005, energy incentives in the US tax code focused on stimulating domestic production of oil and natural gas, at a cost to the government of more than $470 billion. This began to change in 2005, as clearly demonstrated in Tables 3.1 and 3.2 based on Sissine's report.

---

[2]https://www.fas.org/sgp/crs/misc/RS22858.pdf.

Table 3.1 Total DOE energy R&D expenditures for three different periods (billions of 2013 USD)

Technology	FY1948–FY2014 (67-year period)	FY1978–FY2014 (37-year period)	FY2005–FY2014 (10-year period)
Renewable energy	23.0	22.1	7.9
Energy efficiency	19.1	19.7	6.7
Fossil energy	49.3	33.9	10.0
Nuclear energy	97.4	50.1	11.7
Electric systems	8.7	8.9	6.3
Total	204.0	132.7	42.5

Table 3.2 Total DOE energy R&D expenditures as a share (%) of funding (based on Table 3.1)

Technology	FY1948–FY2014 (67-year period)	FY1978–FY2014 (37-year period)	FY2005–FY2014 (10-year period)
Renewable energy	12.1	16.7	18.5
Energy efficiency	10.1	14.9	15.8
Fossil energy	24.6	25.6	23.5
Nuclear energy	48.8	37.8	27.4
Electric systems*	4.4	6.7	14.7
Total	100.0%	100.0%	100.0%

*Transmission and distribution (T&D) systems.

Several conclusions can be drawn: Almost three quarters of post WWII incentives have been focused on fossil and nuclear energy. Since its establishment in FY1978, almost two thirds of the DOE's R&D expenditures have gone to fossil and nuclear, approximately twice as much as has gone to efficiency and renewables. In recent years this balance has changed, and in the decade since 2005 this ratio was 50.9/34.3 = 1.48.

Incentives for improved energy efficiency and renewable energy (largely solar, wind and biofuels) were introduced in FY2006 and by FY2011 accounted for 78% of a substantially increased amount of federal energy-related tax expenditures in that year. What accounted for this temporary spike in support for clean energy was passage of the American Recovery and Reinvestment Act of 2009 (ARRA), which injected an estimated

$20.5 billion in tax preferences into the tax code in FY2011. An additional $3.4 billion was provided in FY2012 by the DOE in R&D support for fossil fuels, nuclear energy, energy efficiency, and renewable energy. At no time were the previous tax code incentives for fossil and nuclear reduced despite growing pressure to do so. In fact, incentives for fossil fuels have increased in recent years as domestic production of oil and natural gas has increased due to fracking.

3.2.4 What Has All This Meant for Solar PV?

As an emerging energy technology, solar has had to contend with the usual gamut of market barriers and market failures. These include

- commercialization barriers arising from competition with well-established technologies with existing infrastructure;
- price distortions arising from existing subsidies and unequal tax burdens in the tax code;
- limited access to capital—e.g., Master Limited Partnerships (MLP), which is a type of limited partnership that is publicly traded, have been established for the oil and natural gas industries to facilitate investment, but not for solar;
- a number of market failures, ranging from lack of access to capital to lack of information for the public to not including the costs of externalities such as air pollution in setting market prices;
- Resistance from the utility sector, which saw a challenge from distributed generation to its traditional business model.

Beginning in the 1970s, the DOE R&D funding for renewable energy began to increase, with solar receiving the largest support among the renewable electric technologies. Primary R&D targets were improved conversion efficiencies from solar radiation to electricity and reduced manufacturing costs. This support dropped significantly during the Reagan Administration but started to rise again in the George H. W. Bush administration when President Bush designated the Solar Energy Research Institute as NREL, the National Renewable Energy Laboratory. While these R&D investments addressed one part of the market barriers limiting the development of solar, little was happening to educate the public about solar and its benefits, financing of solar projects was hard to come by, and markets refused to acknowledge externality costs.

This situation began to change as the public became increasingly aware of global warming and climate change associated with burning fossil fuels. Economies of scale in manufacturing solar modules were also beginning to be achieved in the United States and several countries in Europe and Asia, experience with solar energy systems was growing, and attention was finally being paid to solar financing, which is discussed in detail in this section of this book. A critical step was the introduction of feed-in tariffs (FiT) for solar in Germany, which although not the sunniest geographic location today leads the world in solar installed capacity. Germany has also provided 10-year interest-free loans from a government bank to stimulate solar installations. Many other countries are now providing FiT and other incentives for the installation of solar and other renewables as well.

A second critical development was China's entry into large-scale manufacture of solar modules, which brought solar PV costs down dramatically—more than 90% since 2008. In the United States, the DOE established the Federal Energy Management Program (FEMP) to help federal agencies take advantage of state incentives for solar and other renewables, as well as DSIRE, the Database of State Incentives for Renewables and Efficiency (http://www.dsire.org) as a comprehensive source of information for consumers on incentives that support renewables and energy efficiency.

In addition to state and local incentives for solar energy systems, there is a 30% federal Investment Tax Credit (ITC) for installations on residential and commercial properties. A tax credit is a dollar-for-dollar reduction in the income taxes that a person or company claiming the credit would otherwise pay the federal government. It is based on the amount of investment in solar property and provides market certainty for companies to develop long-term investments that drive competition and technological innovation, which in turn, lowers costs for consumers. It was recently extended to 2023, with the following provisos: it steps down to 26% in 2020 and 22% in 2021. After 2023, the residential credit will drop to zero while the credit available for installations by utilities and commercial operations will drop to a permanent 10%. Commercial and utility projects that have commenced construction before December 31, 2021, may still qualify for the 30%, 26%, or 22% ITC if they are placed in service before December 31, 2023.

3.2.5 Concluding Remarks

Given the current low cost of solar PV, with prospects for further reductions in the future, why do we still need policies and subsidies to support it? Several answers are possible: Solar and other emerging renewables are competing with technologies that are still receiving subsidies that create an uneven commercial playing field. Removing these legacy and now unnecessary subsidies, which is opposed by many in the current US Congress, would be an important step in letting market competition determine our energy investments. A second answer is that even if solar electricity is finally competitive at the point of sale, it and other renewables are still working with an energy infrastructure— a utility regulatory system, a power grid, a centralized energy system—built for a world powered by fossil fuels and nuclear. This infrastructure was built largely with public funds, whereas new energy sources often have to build their own infrastructure, which impacts their market costs and competitive position.

Nevertheless, the transition away from dependence on fossil fuels and toward an energy system increasingly powered by renewable energy is clearly under way. Our responsibility is to make this transition occur as quickly and smoothly as possible.

3.3   Wall Street and Financing

Michael Eckhart[3]

(The Glossary for Chapters 3.3 to 3.7 is on page 158–159)

3.3.1   Policy Drivers for Solar Energy Financing

Financing is generally arranged with two types of funds: equity for the at-risk elements of a situation, and debt where the likelihood of repayment is less risky or reasonably assured. Money is "priced" according to the perceived risk on a scale from low to high:

- For US Treasury Bonds, the risk of default is deemed to be zero, and the interest rates are accordingly very low. Today, Treasury bonds yield about 0.5% for 30-day bills, to 1.75% for 10-year treasury bonds, to about 2.5% for 30-year bonds.
- For real estate loans, the risk used to be considered low due to continuously appreciating property values. After recovery from the financial crisis of 2008, it is still deemed to entail relatively low risk and hence interest rates are low. Interest rates on a 30-year mortgage are typically about 1.75% above 10-year Treasury Bonds for credit-worthy borrower, or about 3.5% for borrowers with a high credit score, and as much as 8% to 12% for those with a poor credit score.
- For corporate and municipal debt, the risk ranges from low to high, and the world uses "credit ratings" to convey the perceived risk, from AAA for lowest risk to D for very high risk. Debt is priced in accordance with the ratings. A rating of BBB- is deemed to be the lowest level of "investment grade risk." Today, Triple-A rated debt yields about 2% to 3%, while a BBB-rated debt would yields about 5% to 7%.
- The most expensive form of debt is in a category called junk bonds, which are unsecured and hence have equity-like risk levels and are called "High Yield" debt. This category of debt yields about 7% to 12% today.
- Preferred equity is the senior-most level of equity capital, typically having a promised dividend but also sometimes having an equity upside, too. Preferred equity returns across a wide range, but often in the 10% to 15% range.

---

[3]Michael Eckhart's biography is on page 291.

- Common equity is the core equity capital of a company or project. Its value rises and falls with the success of the company or project. The equity investor is the "last to get paid" after the issuer covers all of the other sources of money more senior to it (as described above). The investor can receive dividends while owning the shares plus any gain from selling the investment to another party—either privately or publically through a stock market. Equity investors often look for a potential return of 10% to 20%, but sometimes accept a lower or seek a higher level.

- Private equity: A special sub-category of equity capital is called private equity, typically in the form of dedicated funds for institutional investors. Fund managers then on-invest the money into the equity of privately held companies that they often manage for a maximum near-term profit. The average returns of private equity funds were running 3% to 5% above public stock market returns such as the Standard & Poor's 500 Index during the 1990s and 2000s but are reported to have declined to just match the S&P 500 on average today.

- Venture Capital: A subset of private equity is the venture capital industry. The VCs then on-invest the money into seed-stage and early-stage new ventures. The VC system as we know it today developed its modus operandi by investing in technology industries like electronics, computers, software, telecommunications, the Internet, and the Cloud, plus other fast-growth businesses like genetic engineering and pharmaceuticals. Many did exceedingly well, generating returns of 20%, 40%, and more. However, the VC industry did far less well with its investments in solar and other renewable energy, for three reasons: first, the amount of capital required to commercialize a new energy technology was greater than their funds could support. Second, the gross margins in energy technology businesses are far lower (maybe 20% in energy versus 80% in software) so growth could not be self-funding and this created the need for continuing infusions of investor capital. Third, the time required for market adoption of new energy technologies exceeded the typical 10-year term of a typical VC fund.

In sum, the cost of capital ranges from low single digits to well over 40% internal rate of return (IRR). The pricing of capital depends on the opportunity and the perceived risk. In the energy space, many of those factors are established by public policy, as further described in this chapter.

3.3.1.1 The importance of policy to financing

Financing occurs in the presence of all conditions that could affect it. That is, financing considers all of the economic conditions and trends, social circumstances whether war or peace, government policies no matter whether positive or negative, and market conditions including public and private markets, and debt and equity markets, as well as the particulars of the deal at hand.

Financing takes into account two key factors: return and risk. Return can be calculated, but risk can only be assessed. There is always a "base case" of numbers that express the "expected" returns based on contracted revenues and cost, from which one can adjust the numbers up and down to create best case and worse case scenarios, but these are really quantifications of human assessments.

It is said that clients are in the business of creating returns, while investors and bankers are in the business of assessing risk and, on that basis, "pricing" the deals.

Policy can increase risk if they create uncertainties, such as Presidential elections every four years, or if they threaten to add costs, such as regulations. The fundamental policy risk is the risk of change that could alter the attractiveness of a deal after it has been made. This was the case, for example, when Spain stopped its Feed-In Tariff and changed the prices on existing projects retroactively.

Policies also can improve returns and reduce risks. Examples include the 1935 Federal Power Act (FPA) that stabilized the electric utility industry in the form of regulated monopolies, the 1978 Public Utility Regulatory Policies Act (PURPA) that established the legal framework for Independent Power Producers (IPPs), and the Federal Energy Regulatory Agency's (FERC's) Order 888 that gave all players open access to the electric transmission system.

Alternatively, policies can improve returns and reduce risks temporarily such as the 30% Investment Tax Credit (ITC) for solar energy that was scheduled to expire at the end of 2016 but then was extended at the end of 2015, now scheduled to ramp down starting in 2020 and expire in 2021. Another example of temporary impact was the package of economic recovery tools in 2009 that included a special research and development budget, loan guarantees, and cash grants in lieu of tax credits for renewable energy projects and companies.

Therefore, policy is not only important but also the legal framework within which financing occurs. For solar energy in the United States, there are critical policy elements at the federal and state levels.

3.3.2 Federal Policies

The Federal government does not have direct, controlling mechanisms to drive solar energy. Instead, it affects the course of things indirectly by funding research, development, and demonstration (RD&D) programs for new solar and other clean energy technologies, establishes the legal basis for nonutility power generation, provides incentives and financing support for commercialization and initial market growth, and purchases solar energy for its own use, as described in this section.

3.3.2.1 Federal RD&D

The DOE, NASA, DOD, and National Labs like the National Renewable Energy Laboratory (NREL) conduct RD&D and fund companies (contractors) to do the same. Indeed, solar PV technology is a result of government-funded research going back to Edmond Becquerel's first observation of the photoelectric effect in 1838, Albert Einstein's mathematical explanation of the photovoltaic effect (which won the Nobel Prize in 1922), funding of Bell Labs which produced the first practical solar cell in 1953, and the Naval Research Laboratory's Project Vanguard funding its first practical use in 1958.

As described in Chapter 3.1, following the oil crisis of 1973, then President Gerald Ford established the Energy Research & Development Administration (ERDA) in 1975, including a PV program.[4] This was vastly expanded in 1977 when ERDA, the

---

[4]A detailed description about the U.S. Government's involvement in renewable energy, including PV, can be found here: Allan R. Hoffman (2016). *The U.S. Government and Renewable energy: A Winding Road*, Pan Stanford Publishing, Singapore.

NASA solar program, and others were merged in the formation of the US DOE in 1977, with the PV program then managed by the pioneer Paul Maycock.

The Solar Energy Research Institute (SERI) was established by President Jimmy Carter in 1979, initially headed by Denis Hayes (who also founded Earth Day in 1972), and then recast by President George H. W. Bush as the National Renewable Energy Laboratory (NREL) in 1989.

Other of the national labs were drawn into the program, especially the Jet Propulsion Laboratory (JPL), which established the most important government program—the PV quality management program under the direction of John V. Goldsmith—the result of which was that PV module manufacturers are able to provide a 25-year guaranty on the performance of their products. Another important laboratory was Sandia National Labs. Government-funded R&D spread globally, initially in Japan and in Europe, which was managed by the great European PV leader Wolfgang Palz, who went on to write the landmark White Paper on Renewable Energies in 1997, a document that launched Europe's commitment to renewables and continues to be the basis of policy thinking to this day.

There was a sense of competition but also great collaboration among the researchers around the world, such that the technology was uniformly developed. Indeed, it can be said that PV is the only electrical device that is produced everywhere to a single compatible set of standards. A PV module made anywhere can be installed anywhere. This is absolutely key to today's massive financing of PV projects. Of course, quality must be managed and reviewed, but it also can be insured against technical risk or backed by performance guarantees.

It is an interesting anecdote about how the PV adopted a standard 25-year performance guarantee. It was at first a marketing gimmick that Siemens Solar announced in 1999 when it achieved a cumulative total of 100 MW produced! The announcement was made at a national PV conference in Palm Springs, California, that was held outdoors under the lights (that were powered by a diesel gen-set running in the background!).Recall that the Siemens factory was in Camarillo, California, originally built by ARCO Solar and later owned by Shell Solar and Solar World.

This caused quite a stir in the industry and there was a global effort to run lifetime testing of PV panels. This became the forte of Sandia Labs, among others. Then word began to leak out that Sharp Solar in Japan had run their own tests and determined that their standard module would last, in a nonfreezing environment, as much as 125 years! Again, this kind of news had a tremendous effect on the early financiers of PV installations.

One way to view government funding of RD&D is advancing science, and another way is to see it as a stand-in for venture capital. Spire Corp was the leading ERDA/DOE contractor for developing what is today the world standard equipment for PV module manufacturing. Allen Barnett, founder and CEO of Astropower, once remarked that 38% of the company's funding to that date was from DOE RD&D funding, while 62% had come from investors.

All of this made a tremendous shift in the 2004–2009 era when Silicon Valley-style venture capital came into the PV industry in force, seemingly funding anyone who had a solar proposal. For the most part, that phase went bust. In the view of some, it was because of unrealistic expectations for quick R&D success, ignoring the fact that that it takes years and decades to develop a PV technology and its manufacturing processes in a market where there was no opportunity for selling at initial high prices to "early adopters" and then driving costs and prices down over time. The drive on cost reduction was already under way by the time the VCs arrived.

What the Western PV industry did not see coming was the Chinese PV industry, which gained easy access to PV cell technology and manufacturing process technology from the Europeans, plus plentiful cheap capital from the Chinese banks, and already enjoyed a manufacturing culture well suited to semiconductor products. Although the Chinese drove PV cost and pricing down furiously in the 2006–2014 timeframe, quality did not decline because they were deploying European process equipment. The only surviving Western PV manufacturers were First Solar and SunPower, both of which located their major manufacturing facilities in Asia. SunPower was founded by Richard Swanson, a Stanford professor who invented back contacts for solar cells. The company was built up successfully by CEO Tom Werner, who came out of the semiconductor manufacturing industry.

The government-funded important R&D projects helped to build and strengthen the terrestrial PV industry by sponsoring demonstration projects and developing high quality PV modules. The net result today is that the world has plentiful supply of first-tier, high-quality PV modules (described in Chapter 2.6) and associated equipment that the financial community has accepted and ruled as low risk.

3.3.2.2 Public Utility Regulatory Policies Act

In the mid-1970s, there was great expectation about the development and deployment of new energy technologies, especially in the electric utility sector. These included solar, wind, small hydro, geothermal, biomass, and ocean sources of energy plus related new technologies such as storage, fuel cells, energy efficiency, and others. However, the electric utilities were resistant and established their own R&D institution in the form of the Electric Power Research Institute (EPRI) that they funded themselves. The EPRI had research programs in all of the renewable energy technologies, but the industry was decidedly resistant to immediate or near-term deployment while they were dedicated to preserving the nuclear power and coal-fired power industries.

There was an example, however, of a first-of-a-kind nonutility generator. It was the Lowell Hydro Project, developed by J. Makowski Company in Lowell, Massachusetts. It was accomplished under long-standing state law that allowed property owners—presumably farmers—to deploy small hydro in dams and required the utilities to buy the power. What was different about Lowell was that it was a utility scale at over 10 MW and was developed not by a farmer but by a company that said it was in the nonutility power business. The news of Lowell reached Washington at the very time that a debate was heating up about the need for new competition in the utility business. Lowell was a model for PURPA (described in Chapter 2.1), which ordered the utilities to interconnect nonutility sources of power and pay for the electricity at their "avoided cost." That is, at the cost they would be incurring from their next best choice internally.

PURPA put limits on the scale of the new NUGs: 30 MW for wind, solar and other "alternative energy" sources, and 80 MW for co-generation (today generally called combined heat and power after several "tri-gen" ventures were formed. The constitutionality of PURPA was tested in the Supreme Court, which ruled in its favor in 1983 largely because wholesale electricity flows across state lines and is therefore "interstate commerce."

PURPA also had a limit of 49% ownership by utilities. This rule was set aside in the Energy Policy Act of 1992, allowing unlimited utility ownership. This was allowed because another ruling had already put limits on the utilities—not on ownership but on to whom they can sell the power.

It was a case involving Southern California Edison (SCE) and its IPP sister company Mission Energy. Mission was owning Non-Utility Generations (NUG) at the 49% level but in reality controlling the projects, and was entering power sales agreements with SCE. The industry joke was that the SCE and Mission Energy staff sat together in the same offices, but claimed to be separate arms-length businesses. The California Public Utility Commission ruled in 1986 that this practice was forbidden. This created the now-standard practice that nonregulated utility affiliates can own projects anywhere except in the sister company's service territory and cannot sell power to its sister company in any case. It is noteworthy that Mission Energy went bankrupt in 2012 and its assets were sold to NRG.

It is generally considered around Washington, DC, that PURPA was written by Adam Wenner at the law firm of the Sutcliffe law firm, now merged into Orrick. Of course, there were hundreds of other co-authors because it has been so successful.

3.3.2.3 Investment tax credits

The solar ITC was first enacted in 1979 under President Jimmy Carter and then was allowed to sunset in 1983 under President Ronald Reagan. It was reinstated under President George W. Bush in the Energy Policy Act of 2005 and then extended for an eight-year period in the Economic Restoration and Recovery Act of 2008 also under President Bush.

Interestingly, most Americans thought George W. Bush was pro-oil and anti-renewables, but this was untrue. In fact, while he had delegated energy policy to his Vice President when they took office in January 2001, he then came back to it in his final year in office, giving a major energy speech including renewable energy at the Washington International Renewable Energy

Conference (WIREC) in March 2008 that was produced by the American Council On Renewable Energy (ACORE) working jointly with the Bush White House Council on Environmental Quality, then headed by James Connaughton and the State Department under the supervision of Secretary of State Condeleeza Rice. This global meeting of some 8,600 people from around the world (1,100 government officials and 7,500 people from industry and civil society) took place just six months before President Bush extended the 30% ITC for eight years in September 2008.

The solar ITC became a cause celeb in 2014–2015 as it entered its final two years, scheduled to sunset on December 31, 2016. What would happen if the ITC was not extended? What would happen to utility-scale solar and to residential (resi)-rooftop solar? The message to Congress was PLEASE give us an extension, while the message to investors and lenders was "not to worry, we have a plan and we'll do just fine." At the end of 2015, the situation was dire. There was great concern that the ITC would be allowed to expire, sending the US solar industry off a cliff into a valley of unknowable depth.

Then, without notice or warning, the Senate passed a permanent extension of the solar ITC and wind PTC in the form of ramp-downs. What had happened? As this author learned, Senator Harry Reid and his staff had done it themselves. But why and how did it happen? It was months later that this author heard, at a meeting of the oil and gas industries, that they had won a reversal of the decades-long prohibition of exporting crude oil in a political deal—in exchange for allowing the Congress to extend the wind and solar tax credits! It was a deal cut one evening in December 2015, when the oil industry called on Senator Reid and asked, "What will it take to get you and the Democrats to support this?" Senator Reid reportedly replied, "We want an extension of the wind and solar tax credits," to which the oil representatives reportedly said: "Agreed, if you give us your language tonight." Whereupon Senator Reid and staff wrote it up quickly and sent it over—and it was done.

Incidentally, the Wind Power Production Tax Credit was first enacted in 1979 and terminated in 1983 but then put back in place in 1992 with the leadership of Senator Bennett Johnston, Jr., of Louisiana, who was then the Chairman of the US Senate Energy & Natural Resources Committee, and who said at the time that he wanted a tax credit and not a cash subsidy so that the entrepreneurs who develop the wind farms would have to sell them to large corporations (utility companies) to own and run them, saying that "if we have to have wind power then I want it owned by responsible adult tax-paying corporations."

This was a companion to Title 12 of the same act that eliminated the utilities' 49% ownership limit of IPP projects, then allowing them unlimited ownership as Exempt Wholesale Generators (EWGs). This fundamental change in the structure of the utility industry immediately led to the creation of utility holding companies and nonregulated generation companies as sister companies to the regulated utilities. Examples include NextEra with Florida Power & Light and NextEra Energy Resources, Sempra with San Diego Gas & Electric and Sempra Energy Resources, and Exelon with Commonwealth Edison and Exelon Generation.

While PURPA established the IPP industry in 1978, it was the refinements in the Energy Policy Act of 1992 that molded the industry into the form it is today. These laws are the basis for the financing of solar and other renewables today.

3.3.2.4 Commercialization and deployment

The DOE provides loan guarantees for renewable energy, nuclear power, advanced fossil fuel technologies, and clean energy manufacturing. There is an exhaustive (and bidders say exhausting) process of project evaluation and selection, and the government naturally wants to avoid taking risk (even though the very purpose of the program is for the government to take on risks that the commercial sector will not take) in situations where there can be a national benefit if it all works well.

The DOE loan guarantee programs are of two types: Section 1705, enacted in 2005, provides loan guarantees for "advanced and innovative" technologies for which the recipient pays the risk reserve fund, and Section 1707, enacted in 2009, which provides loan guarantees for "shovel ready" new technologies for which the government pays the risk reserve fund.

It was not surprising that no projects were funded prior to 2009 while the DOE had one person assigned to administer the program, while it boomed in the 2009–2011 period when (a) the DOE staffed up the program with true financial professionals under program director Jonathan Silver and (b) the industry was ready with exceedingly large-scale PV, CSP, and wind power projects.

One example was the 550 MW Desert Sunlight project developed by First Solar. The project, costing $2.3 billion, was structured with $600 million of equity, provided 50/50 by GE Energy Financial Services and NextEra Energy Resources (Sumitomo acquired an ownership interest later), and $1.7 billion of debt under the advice, structuring, placement and administration by Goldman Sachs and Citigroup. The debt was divided approximately into $750 million of project bonds, $650 million of syndicated bank loans, and a $300 million three-year loan to the federal cash grant (taken in lieu of the ITC). The project was pulled forward and made possible by Power Purchase Agreements (PPAs) (see also chapter 4.3.2) from Southern California Edison (SCE) and Pacific Gas & Electric (PGE), acting under the requirements of California's RPS, and provided the credit worthiness for the nonrecourse financing provided to the project. So this was a case in which federal and state policies dovetailed beautifully to create a commercial success supported by policy.

There were five other utility-scale solar PV projects financed with the federal loan guarantees:

- Aqua Caliente, owned by NRG and Mid-American
- Alamosa, owned by Cogentrix and Carlyle
- Antelope Valley, owned by Exelon
- California Valley Ranch, owned by NRG
- Mesquite, owned by Sempra and Con Ed.

The eventual effect of the DOE loan guarantee program was to successfully demonstrate to investors and lenders that utility-scale PV was sufficiently economic for the utilities, sufficiently profitable for the investors, and sufficiently low risk for the lenders to go forward.

In addition, there were four utility-scale CSP projects supported by the federal loan guarantees:

- Crescent Dunes, owned by Solar Reserve and Santander
- Genesis, owned by NextEra
- Ivanpah, owned by Brightsource, NRG and Google
- Mohave, owned by Atlantica Yield/Abengoa.

However, CSP was proven to be too high in cost per kilowatt-hour to be competitive going forward. While PV was being bid in 2016 at 3 to 6 cents per kWh, CSP was stuck at a price range of 12 to 15 cents. Great pioneers like John Woolard at Brightsource and Kevin Smith at SolarReserve did valiant work to put CSP forward.

What one saw for PV was a perfect combination of factors as

- China ramped up PV manufacturing in the period 2008–2012, driving down costs and prices of panels by some 75%;
- Interest rates declined in the 2012–2016 period to historically low levels;
- Industry gained experience in building utility-scale projects thus driving down total installed costs.

Everything has gone right in recent years for PV, but CSP did not have enough cost reduction opportunity from scaling up manufacturing. PV is a semiconductor for which there is a well-evidenced 18% to 20% learning curve, while CSP is made of glass and metal for which there is said to be a 3% to 6% learning curve.

It was interesting that the loan guarantees were also extended to clean energy manufacturing companies, which did not work out as well, as there were defaults by Solyndra making solar PV, 123 making batteries for energy storage, Fisker making an electric vehicle, and Beacon Power making flywheel energy storage.

Overall, however, the DOE loan guarantee program has been deemed a tremendous success, both as to financial performance and impact of the commercialization of the technologies especially solar PV and wind power, and also EVs by Tesla.

3.3.2.5 Government purchasing

In some sectors, the government has become a large single customer. This is the case for energy efficiency and increasingly for solar energy. In addition to federal and state government purchasing, there is a quasi-government market called the MUSH market (municipals, universities, schools, and hospitals) that is a big market for solar energy and energy efficiency.

However, while big, it is difficult to serve because it is so spread out, hard to sell to, and typically buys from the low-price bidder—so it is hard to make a profit. However, on the other hand, this is common for all government procurement, and a company must specialize in this market to be successful. Many are doing so.

Government purchasing of energy efficiency retrofits under a nationally uniform Energy Savings Performance Contract (ESPC) for federal government facilities is a big success. The uniformity of the contracts allows for securitization of the contract cash flows. This activity has been dominated by Hannon Armstrong, a small investment banking firm in Annapolis, Maryland, that has subsequently gone public as Real Estate Investment Trust (REIT).

3.3.3 State and Local Policies

There is well known saying that "all development is local." While there are many national and international development firms, they must have regional or local management and professionals who are knowledgeable about local laws, regulations, procedures, practices, and culture to be successful. This section summarizes some of the key public policies that affect the successful development and operations of IPP wind, solar and other renewables.

3.3.3.1 Renewable Portfolio Standards (RPS) and RECs

Many states, municipalities (and countries) have rules that specify that utilities must have $x\%$ of their energy supply from renewable energy sources by a target date. In the United States, these are called Renewable Portfolio Standards (RPSs) (see also Chapter 4.3).

3.3.3.2 Solar Set-Asides and SRECS (see also Chapter 4.3)

In a situation where a government has an RPS in place, but no other rules, most if not all of the winning bids in a procurement will go to wind power because wind has come down the learning curve first, and its costs have decreased ahead of solar end other renewables.

Solar energy advocates successfully lobbied to have solar set-asides put into the RPSs. For example, such an RPS would require 20% renewables by a certain year, with not less than 20% of it from solar.

The RECs created under a RPS with solar set-aside are called SRECs. In the early years of a solar set-aside with SRECs, the requirement can be greater than the industry can supply, creating a shortage and attendant high SREC prices. The most extreme example was in the state of New Jersey where the electricity from a solar PV system might sell for $10/MWh, while the SRECs were selling for $50-$60/MWh. The problem with RECs and SRECs is that they are issued on a one-year basis, so they are not considered by senior lenders that need to see multi-year contracted revenues.

3.3.3.3 Net energy metering

Net energy metering (NEM) allows a utility customer, whether residential, commercial, institutional, or industrial, to generate electricity at its own cost on its side of the meter and thereby offset its retail purchases from the utility—more or less like growing tomatoes in your back yard and not buying them from the grocery store, except in this case the grocery store is monitoring what you are doing and is displeased. In effect, the customer is "selling" solar electricity at the retail rate, which averages 12 cents in the US and can be as high as 35 cents per kWh in Hawaii and other locations. This is much more than prevailing wholesale generation rates that apply to IPP projects. The person most responsible for the national adoption of NEM was Tom Starrs, who, then an independent attorney, went state by state in the late 1990s and early 2000s, lobbying for NEM.

The combination of NEM, declining PV system costs, very low interest rates, and a 30% ITC in the United States come together to create a very profitable situation. These same factors apply in varying degrees around the world.

However, there is increasing pushback on NEM from the utilities and their regulatory commissions. This is resulting in fixed charges being made to NEM customers to "pay for the grid" and reconsiderations of the rate being paid, looking at negotiated rates rather than full retail offset. As this book goes to press, the future of NEM is uncertain and will be a major factor in the future of customer sited rooftop solar.

3.3.3.4 Leading state examples

Each state has established a unique program of policy for solar and renewables. Some of the leading states are the following:

California: This is the leading state for solar with a rich package of finance-relevant policies including a cash subsidy in the 1998–2008 era, a robust RPS that has risen from 25% to 33% and presently 50% by 2030, a strong full retail off-set NEM, and state-wide permitting and building code for PV.

New Jersey was the second boom state for PV using a strong RPS requirement and solar set-aside within the RPS and SRECS that rose to very high levels. A major utility, Public Service Gas & Electric, has implemented a solar PV financing program.

Massachusetts has been an excellent location for PV, much like New Jersey with a strong RPS, a solar set-aside and SRECS. As this book is written, it appears that the policies might be reversed.

North Carolina distinguished itself early on with its North Carolina Solar Energy Center and state tax credits.

New York has had the New York Energy Research & Development Agency (NYSERDA) for many years and more recently has established the NY Green Bank and is embarking on fundamental reform of the electric utility sector under a program called Renewable Energy Vision (REV).

Vermont is the first state to establish an RPS for 100% renewables, undoubtedly to be copied by other states over time.

3.3.4 International Policy for Solar Energy Financing

The story of solar investment and finance would not be complete without considering that tremendous influence of government policy around the world that has and is affecting the course of solar energy financing as a driver of market growth. There are five parts to this topic:

- policies of individual governments such as the United States, Germany, Europe as a whole, Japan, China, and "Rest of World"
- the policy push of international agencies such as IRENA, REN 21, the IEA, the OECD, and, of course, the UNFCCC
- the multi-lateral development banks (MDBs) such as the World Bank/IFC, EIB, EBRD, ADB and others
- non-government organizations (NGO)
- private sector players in the policy arena

3.3.4.1 Policies of individual governments

In the mid-1990s, government policy began shifting from a sole focus on RD&D to a new second focus on adopting solar energy at scale sufficient to change the energy supply, environmental damage, and climate emissions of the world. It was quickly understood that the new policy would need to focus on market acceptance, capacity building, and financing. Here are several examples of leading countries:

Europe as a whole: Many countries can now take credit for the success of renewable energy, but in the author's view the credit goes to the United States for funding the development of the technologies, and to Europe for leading the early adoption of renewable energy at scale. The father of renewable energy policy in Europe was Dr. Wolfgang Palz, who led the renewables program in the European Commission for many years. In the mid-1990s, Dr. Palz was accused of wrongdoing by the nuclear energy advocates, and after a two-year investigation was found to be the only person on earth who has been proven beyond a shadow of a doubt to be 100% honest. Dr. Palz was utterly cleared. During the two-year investigation, Dr. Palz wrote a paper, which was published in 1997 as the "White Paper on Renewable Energies." This became the blueprint for Europe's rapid commitment to renewables, which culminated after several steps in the Renewable Energy Plans of 2004 and 2009. Dr. Palz worked closely with Dr. Hermann Scheer, member of the German Bundestag and leading advocate of solar energy in Germany and all of Europe. Dr. Scheer led a nonprofit organization called Eurosolar, managed by his wife Irm Pontenagel. In 2002, Dr. Scheer and the author called for a first global meeting of governments on the topic of renewable energy only, not sustainability and larger topics which was prevalent at the time. Dr. Scheer gave the idea to then Chancellor Gerhard Schroeder, who made it the topic of his keynote speech at the World Sustainability Conference in Johannesburg that year, which led to the first such meeting—the Bonn Renewables 2004 Conference in June 2004 in Bonn, to which 154 countries sent representatives. This was the breakthrough meeting that launched solar and renewables around the world. The three products of the Bonn meeting were:

- IRENA: the International Renewable Energy Agency, now operating successfully from its headquarters in Abu Dhabi (described in Chapter 4.1);
- REN 21: the Renewable Energy Network for the 21st Century, now operating successfully from its headquarters in Paris;
- Country Action Plans, which became the basis for the Indicated National Development Commitments (INDCs) of the Paris Climate Agreement in 2015.

There were many people who drove renewable energy policy and financing in Europe, including, but not limited to, John Bonda,[5] Hans Joseph Fell, Harry Lehman, Christine Lins, Wolfgang Palz, Hermann Scheer, Arthouros Zervos, and so many others.

Examples of several key countries:

Germany: The early efforts to design policy to drive financing for solar energy occurred in the early mid-1990s in the city of Aachen, where a special price was paid for solar electricity (the "Aachen model"). This was accomplished by Green Party government leader Hans-Josef Fell, who went on the national Bundestag and there was the principal author of the FiT in its original form for solar PV in 1998 and subsequently updated and expanded until the full version for all renewables in 2004. The FiT was designed, not as a subsidy as seen today, but as a legal structure to attract long-term debt capital to PV and other renewables. Indeed, it was based in large part on PURPA—an order on the utilities that they must interconnect and buy the electricity. The pricing formula was based on the US methods for nuclear power pricing. The prices were set by estimating the Revenue Requirements a PV project in Germany needed. Of course there was a range. The decision of the planning group was not to select the lowest possible price (so only one project gets built) but the mid-point of the distribution curve so that the best half of all projects will get built. One early change was to increase the term of the agreement from a five-year period to 20 years, to support longer-term, lower-cost debt.

There is an interesting anecdote about how this was accomplished. One day after a Eurosolar's annual conference

---

[5]John Bonda was the general secretary of the European Photovoltaic Industry Association.

on solar energy, Dr. Scheer asked the author: "How can I get your Wall Street to finance my solar revolutions?"

I replied: "Change the feed-in-tariff from five years to 20 years, and then stand back and watch the capital flow. It is what lenders want."

He asked: "That easy?"

I said: "Yes, that easy."

He said: "Come with me. We will visit KfW (Kreditanstalt für Wiederaufbau ["Reconstruction Credit Institute"], the national development bank of Germany)."

We walked down the street to KfW headquarters (this was before the German government moved back to Berlin and the KfW headquarters back to Frankfurt), and into the office of the President. They spoke in German. Then the President said in English: "Yes, Dr. Eckhart is correct." Whereupon they agreed that, if Dr. Scheer could get the FiT extended to 20 years, the President of KfW would launch, on his own authority, a solar loan program for all Germans. Dr. Scheer reported several weeks later that he had done it in the most remarkable way. He had introduced an amendment to another piece of legislation that was sure to pass. His amendment said: "in previous law x, line y, change the number 5 to the number 20." And it passed. So the German legislators had passed perhaps the most important change in the history of solar energy financing, and the legislators did not even know what they were voting for! And KfW announced its solar loan program soon afterwards, with 10-years loans having an interest rate of 1.75% (in an era when interest rates were normally 6% to 8%).

Japan: This country was the first to embark on PV manufacturing an adoption starting in 1996 with the "Sunshine" program. The early leading manufacturers were Sharp and Kyocera, with others being Sanyo, Sumitomo, and Panasonic. The government offered direct cash subsidies to customers who installed PV of their homes, and local banks offered loans for the installations.

China: The biggest boom in PV manufacturing was in China, starting with passage of the Renewable Energy Law in 2006, which was written by Li Junfeng, who was Deputy Director of the Energy Research Institute with the NDRC and then was directed to organize and lead a new China Renewable Energy Industry Association (CREIA). Keep in mind that industry associations in China are not organized by industry to lobby government—that was not allowed. No, associations were created by government as a way of giving directions to industry, so they are headed by government employees as a side job. But China embarked on PV as an export industry, with essentially no domestic market. This changed suddenly in 2010 when the PV manufacturing industry was vastly over capacity and they needed a domestic market to survive. So a new directive came out, calling for the creation of a domestic market with banks ordered to loan funds to the projects. Within five years, China was installing 18 GW per year, worth roughly $36 billion. It was government orders on the banks that created the PV manufacturing industry, and then government orders on the banks to fund PV installations that created the market. China had become one very big "green bank."

Rest of World: Although much of the early policy innovation and financing took place in the countries above, there have been financing innovations in many countries:

In South Africa, a national procurement didn't just award PPAs in hopes of getting financing, they required financing to be in place just to compete. This drew in the South African banks including Standard Bank, Investec, and Nedbank, along with institutional investors like Old Mutual (insurance company) for $10 billion of financings in just three years.

In Dubai, the Dubai Energy & Water Authority (DEWA) has run a procurement program for some 3,000 MW of PV in which ACWA Power, a Saudi Arabia based company under the leadership of CEO Paddy Padmanathan, won the first 200 MW with a bid of 5.84 cents (the first under 6 cents in the world!), and then Masdar, a UAE-based company under the leadership of Dr. Sultan Al Jaber, won the next procurement of 800 MW with a bid of 2.99 cents just 18 months later. This truly illustrated the new geo-political nature of solar power in the developing world.

3.3.4.2 International agencies

Starting with the all-important Bonn Renewables 2004 conference, there began a number of major developments of international agencies focused on the adoption of renewable energy:

IRENA[6]: The United Nation's International Renewable Energy Agency (IRENA) was the original idea of Dr. Hermann Scheer of Germany, who first presented it to a US Senate hearing on climate change in 1990, chaired by then-Senator John Kerry. Dr. Scheer promoted the idea to establish a UN agency for Renewable Energy sources similarly to the UN's International Atomic Energy Agency (IAEA). However, the UN declined to support the idea, and it went ahead under German sponsorship as an independent agency. Finally it was founded on January 29, 2009. Ironically, the US State Department sent a letter to the IRENA founding on January 18, 2009, informing IRENA of the United States' decision not to join. This was just five days before the inauguration of President Barak Obama on January 22, and the appointment of Hillary Clinton as the new Secretary of State the next day. When Secretary Clinton arrived at her desk on January 23, there was a letter on her desk from the author, speaking for ACORE) requesting a reversal of the prior decision. It was reported that Secretary Clinton wrote "Of course, approved!" on it and gave it to her staff for implementation. On the following Monday, January 29, the US government had its representative, a Mr. Brown, at the IRENA founding meeting, and that May became a member. The headquarters of IRENA was established a year later to be in Abu Dhabi, under the leadership of Dr. Sultan Al Jaber, then the founder and head of Masdar. IRENA is emerging as an important factor in the global adoption of renewable energy under the leadership of the initial interim Director-General, Hélène Pelosse, who was followed by Adnan Amin in 2010 as permanent Director General. IRENA presently (2016) has 147 countries as members and is a global success.

IEA: another source of information that is relied upon by investors and financiers is the International Energy Agency which, until IRENA was founded, was stoutly opposed to renewables. Then, under the leadership of Executive Director Maria van der Hoeven, and in the hands of program director Paolo Frankel, did an about face and became a significant source of objective analysis and forecasting of renewable energy.

---

[6]Because IRENA's most dominant area of work is solar energy, it has extreme importance for the future of solar energy. Therefore, it is described in detail in Chapter 4.1.

UNFCCC: The UN Framework Convention on Climate Change was established in 1990 to coordinate the world's efforts to address global warming and attendant climate change. In a way, it has been both a positive and negative force on the financing of solar energy by driving the world's attention to climate change but also causing a sharp divide between liberals and conservatives in the financial community. However, the positives kicked in stronger after the passage of the Paris Climate Agreement, under the great leadership of Secretary-General Ban Ki-moon and UNFCCC's Christiana Figueres, whereupon the existence of climate "deniers" and climate "doubters" seemed to diminish.

A contributing causation of the diminishment of the deniers was the efforts of state attorneys general to investigate ExxonMobil and other fossil fuel companies on charges of funding intentional misinformation campaigns to shed doubt on the credibility of climate scientists and the fact that human activity has caused global warming and climate change. One such investigation of ExxonMobil is said to have caused the company to come forward with a proposal for the US to enact a revenue-neutral tax on carbon.

3.3.4.3   Multi-lateral development banks

World Bank/IFC: In January 1996, World Bank President Jim Wolfensohn and Rockefeller President Peter Goldmark had a dinner where they agreed to form a joint working group to develop a plan for financing solar energy in the developing countries, especially as to off-grid solar. The working group worked for nearly a year and brought forward a recommendation to form the Solar Finance Corporation (an investment fund) and the Solar Development Corporation (a foundation). But they failed to launch. At about the same time, the IFC tried to launch a Renewable Energy and Energy Efficiency Fund (REEEF) that also failed to launch, with rumors of significant internal opposition from advocates of coal-fired power for the developing countries. This started more than a decade of intentional "non-progress" among the multilateral development banks. But all that changed when Jim Yong Kim became President of the World Bank Group and put it on a course to support and finance clean energy. Since then, the World Bank and its private sector unit International Finance Corporation have become major leaders of solar and renewable energy projects around the world. Some of the early staff leaders on solar finance have included Dana Younger, Loretta Schaeffer, and Vikram Widge, among many others.

EIB: Actually, the European Investment Bank would say that it was the first to move ahead on solar and renewable energy financing because it had support from European governments. Evidence would suggest this is true. And EIB has a way of co-financing with the private sector that can be a model for others. Investment officer Chris Knowles has been a strong leader in EIB's clean energy investments and lending.

ADB: The Asian Development Bank has likewise been a leader among the MDBs on clean energy and especially solar finance in the Asia region. ADB set forward on its own, serving its Asian member countries.

3.3.4.4   Impact of NGOs on government policy

Nonprofit organizations have had a major hand in driving policy on solar financing through thought leadership and convening.

The nonprofit sector includes two principal forms of tax-exempt organizations. One is a trade association, authorized in the US under IRS code section 501(c) (6) to "petition the government" as representative of their respective members, as provided for in the US Constitution. The other is generally called a "non-profit group" or "non-government organization" or NGO, that is authorized under IRS code section 501(c)(3) which covers charitable organizations where education is included as a charitable purpose. NGOs use the educational category to carry out their efforts to change society. 501(c)(3) organizations are permitted to do lobbying so long as it is not more than 5% of their total efforts, or not more than 20% if they file a special Form H that then allows this higher level.

In Europe and many other countries, NGOs are quite prevalent, operating under the laws of each respective country.

It is not unreasonable to say that the global movement of government policy toward renewable energy has been driven by the persistent efforts of the NGOs. Following are several that have had impact:

Eurosolar: This nonprofit organization in Bonn, founded by Dr. Hermann Scheer and managed by his wife Irm Pontenagel, has been a driving force behind the advancement of solar energy in Europe, holding annual conferences including the first to focus on financing as the key to global adoption of solar energy. Dr. Scheer also founded the World Council for Renewable Energy (WCRE), for which Dr. Wolfgang Palz was European chair and Michael Eckhart was US chair.

REN 21: The Renewable Energy Network for the 21st Century also came out of Bonn Renewables 2004. It has become a network of hundreds of renewable energy advocates, policy professionals and researchers around the world who collaborate through the online programs. Early leadership of REN 21 came from Mohamed El Ashry, its first Chairman and previously the first head of the Global Environment Facility (GEF), David Hales, who chaired the Multi-Stakeholder Day at Bonn Renewables, and Griffin Thompson, an energy official in the US State Department, and several members of German government which has funded REN 21 since its formation in 2005. The pinnacle product of REN 21 is its annual Global Status Report on Renewable Energy, which was prepared in its first several years by Eric Martinot, one of the great analysts of the renewable energy community, supported by inputs from REN 21's network of experts. More recently, the GSR has been written and edited by Janet Sawin, coordinating input by over 300 contributors around the world. Information like this is vitally important to investors and financiers who must conduct due diligence analyses of their clients and deals. REN 21 was managed by Virginia Sonntag-O'Brien, and more recently chaired by Arthouros Zervos and managed by Christine Lins, the same two who built EREC so successfully.

ACORE: The American Council on Renewable Energy was founded in 2001 by the author as the US arm of the WCRE, with a philosophy matching what was occurring in Europe at the time. That is, to deal with renewable energies (as they say in Europe) as a whole for the purposes of government policy and financing, while recognizing that trade associations would continue lobbying on a technology-specific basis. ACORE's stated purpose was to bring together all the players necessary for renewable energy to be successful in the United States, and the central mission was to bring renewable energy policy and finance together, in effect, to take renewables away from government and put it into Wall

Street. The aim was to involve the financial community in policy-making, so that policy would shift from its traditional focus on RD&D, and now add Renewable Portfolio Standards, tax credits, accelerate depreciation, and other financing-related topics. ACORE convened the first conference on renewable energy finance in the United States, the Renewable Energy Finance Forum Wall Street, which continues today as a major annual event. ACORE was successful in its mission, which continues today. Key people in the formation of ACORE included Board members Dan Reicher, Michael Ware Ken Locklin, Rob Pratt, and many others. Key staff members who "built" ACORE included Jodie Roussell, Tom Weirich, and Alisa Frederick, plus Dawn Butcher. Later key people included Denny McGinn, Todd Foley, and many others.

EREC: The European Renewable Energy Council filled a role similar to ACORE, except that it was actually a council of all the renewable energy trade associations in Europe. It was organized and led by Chairman Arthouros Zervos and Executive Secretary Christine Lins (who now lead and manage REN 21), who coordinated brilliant analyses of the potential for solar and renewable energy in Europe, providing the analytic foundations for every renewable energy law that was passed by the EU and its member countries in the 2002 to 2012 era.

CREIA: The Chinese Renewable Energy Industry Association emerged in 2006 upon the enactment of China's Renewable Energy Law. CREIA was established by Chinese government to give direction to the solar energy and wind power companies in China, and headed by Li Junfeng, the author of China's Renewable Energy Law.

As some of the influential forces behind the scenes, WCRE, ACORE, EREC, and CREIA, began actively collaborating on advancing finance-oriented policy in 2004 and carried on this "grand conspiracy for good" until about 2011–2012 when the leaders of the groups, Wolfgang Palz at WCRE, Michael Eckhart at ACORE, Arthouros Zervos at EREC, and Li Junfeng at CREIA, all moved on to other positions. It was a golden era of thought leadership, collaboration, and effectiveness, never to be repeated again. This early work of global coordination has now been taken up by IRENA, REN 21, IEA, UNFCCC, OECD, and many other international organizations.

Among the 501(c)(3)[7] NGOs, other examples include, but are not limited to,

- World Resources Institute (WRI);
- Worldwatch Institute;
- Sierra Club;
- The Nature Conservancy;
- Brookings Institute.

And a number of climate-focused organizations, such as

- The Climate Group;
- Carbon Disclosure Project (CDP).
- Business Council for Sustainable Energy (BCSE).
- World Business Council for Sustainable Development (WBCSD).

Among the US trade associations, the major players have included

- American Wind Energy Association (AWEA);
- Solar Energy Industries Association (SEIA);
- California SEIA (CalSEIA);
- Geothermal Energy Association (GEA);
- National Hydropower Association (NHA);
- Many biomass energy and fuels associations.

It is often said that the fossil fuel industries have "hundreds" of lobbyists working to further public policy in their interests, so it is unfair to the renewable energy advocates. However, the reality is that the renewable energy community has "tens of thousands" of advocates at work. This is why renewable energy is winning.

---

[7]The organizations described in section 501(c)(3) of the US Tax Code are commonly referred to as *charitable organizations*.

3.4 Solar Market Segmentation and Financing Methods

Michael Eckhart

This section will focus on the financing of solar PV, but the author wants to acknowledge that there are many segments and sub-segments of the solar business.

3.4.1 Utility-Scale Solar Project Financing

Growth of utility-scale solar: PV and CSP

The DOE PV loan guarantees: One of the great examples of success was the Desert Sunlight Project, seeking up to 80% guarantee of project debt that was up to 80% of capital. The effect was to make large-scale projects financeable that were otherwise not, to extend the term of loans, and reduce interest rates by approximately 50%, which gave a windfall to equity investors thus also drawing the equity capital into the projects.

This experience became the "learning" for banks and other players, such that, by 2014, utility-scale projects were getting done with no guarantees. The program also generated huge demand for PV panels, thus helping to drive down the cost and prices of panels, making PV more competitive. And, electric utility companies have been acquiring ownership positions, creating a profit for the original developers/investors/owners.

3.4.2 Commercial and Institutional Rooftop Financing

There are several ways of segmenting the commercial and Institutional (C&I) market:

- by type of building: MUSH, government, commercial
- by type of contract: one-off or strategic (Wal-Mart, Whole Foods, Amazon, etc.
- by type of financing: end-user (self) financing, lease or PPA

PV on commercial rooftops grew rapidly in the 2010–2015 period with leadership by retailers like Wal-Mart and Whole Foods for which the financing has been provided by third-party leasing and PPAs, and by high-tech companies like Google, Apple, and Microsoft for which financing was done with internal capital.

The challenge in the C&I segment is that there is no nationally applied method or program for financing of PV on roofs where the PV systems are interconnected on the customer-side of the meter. The rules vary by state.

3.4.3 Community Solar

A new market is emerging called Community Solar, in which customers can participate in a shared solar system that is not on their roof but still receives the financial benefits of net energy metering.

In the United States, this began in the state of Colorado by several innovative young companies: SunShare, founded by community solar pioneer David Amster-Olszewski, Clean Energy Collective, and Grid Alternatives, and supported by investors NRG Renew and Excel Energy.

The financing of community solar can be challenging because of the question "who is the credit?" for the purposes of senior debt lending. Certainly, the PV project serves as collateral for the loan, but is the loan based on the credit worthiness of the off-take utility or the owners of the project? Who or what takes responsibility for loan payments during that period? As of 2016, it is still the pioneering stage for community solar and the pioneering companies are developing a range of financial structures.

3.4.4 Residential Rooftop Financing

The financing of residential ("resi") rooftop PV has been nothing short of a revolution started by such pioneers as

- integrated solar installation companies such as SolarCity, Vivint, Sungevity, and others, and
- specialists in solar financing: SunRun, Spruce Capital (formerly Clean Power Finance and KW Financial), Mosaic, Admirals Bank, and others.

The financing of resi rooftop solar is driven by the application of the federal Investment Tax Credit (ITC). There are three prevalent methods: the PPA Model, the leasing model, and the standard loan-to-own model.

3.4.4.1 PPA model

The homeowner enters into a Power Purchase Agreement (PPA) for the purchase of all the electricity produced by the PV system at an agreed-upon rate per kilowatt-hour (kWh) (typically with an annual escalator) for a specified period, typically 20 years. The PV system is sold by the installer to a legal entity called a special purpose vehicle (SPV) that owns the PV system and is itself jointly owned by the installer and a third-party financier (the tax equity investor). The homeowner continues to be billed by the utility for electricity it sells to the homeowner, but the amount has been reduced by the electricity generated by the PV system, hence paying the "net" amount. The homeowner also is billed monthly for the kWhs generated by the PV system under the PPA and pays the SPV owner.

The total amount of the utility bill and the PPA payment should be less that the amount of the utility bill before there was a PV system. The money paid to the SPV is split between the third-party tax equity investor and the installer. At a point in time in the future, when the tax equity investor has recovered its investment plus an agreed-upon rate of return, the economic interests "flip" and the installer is back as owner. This is called the "partnership flip model." At the end of the PPA, the owner can enter into a new PPA and continue on, purchase the system, or require the installer to remove the system at its cost.

Innovation on this approach was begun by pioneers like Greg Rosen, who was the first person in the world having a full-time job financing solar, then at PowerLight in about 2000 (acquired later by SunPower), so Greg was the real pioneer on all of this.

3.4.4.2 Inverted lease

This is similar in some ways to the PPA model in that there are three parties involved: the homeowner, the installer, and a third-party tax-equity investor. In this model, however, the homeowner simply leases the PV system for a fixed monthly payment and allows the PV-generated electricity to reduce the utility bill due to net energy metering. These leases are highly complex in nature because of the tax rules for leasing.

For third-party financing and use of the ITC, blocks of capital (i.e., $50 million, $100 million, even $250 million) are awarded to the installation companies by tax equity investors of which there are only about 20 in the United States. These include major banks such as BAML, CS, JPMC, Citi, and others, plus several high-tech firms that want to invest in clean energy and can use the tax credits such as Google.

The typical approach, called the inverted lease, is a mind-bogglingly complex financing and legal structure that seems "simple" to the homeowner, that was invented by Marshal Salant and Jason Cavalier when they were at Morgan Stanley in 2008. It was a great contribution to the financing of solar energy.

3.4.4.3 Loan-to-ownership

This straightforward approach to financing an installation—as is done as loans for autos and other major purchases—was set aside in the early years of the resi-rooftop PV market because of the existence of Net Energy Metering, the 30% ITC from 2005 to 2022, and exceedingly low interest rates prevailing in the period 2012-present, allowing for 100% leases and "no money down" to buyers. With this financing method, solar PV went from being "too expensive" to being free. Amazing turnaround.

However, in 2015, buyers were becoming more sophisticated and competition was heating up, so that buyers saw the benefit of taking a loan on their solar PV system, even if they have to put down 10% or 20% of the purchase price, because they would personally take the 30% ITC, thus paying back their down payment within a year, then having a lower cost of electricity, and then owning the system and getting free electricity after the loan is paid off, for life of the system.

In 2015, solar companies like SolarCity, SunRun, and others were offering loan programs alongside their lease and PPA programs. And solar finance specialists like Mosaic, led by pioneer Billy Parish, were focusing on this method. A shift was already occurring as the market prepares for the ITC ramping down and expiring.

3.5 Solar Project Financing

Michael Eckhart

In the landmark study entitled "Solar PV Industry, Markets and Technology" prepared in November 1978 by the consulting firm Booz, Allen & Hamilton for the DOE and the Carter White House (with the author as principle investigator) a major headline conclusion was "The role of electric utilities in the commercialization and adoption of solar PV is unclear and needs national policy attention."

To this day, the utility industry has struggled with that question, while the solar and other renewable energy industries have emerged under PURPA and its subsequent laws, with variations on project financing.

3.5.1 Traditional Power Generation Financing

The IPP industry has created quite a transformation in how power generation facilities are financed.

Traditionally, and to some extent today, power generation facilities were financed on the balance sheet of the electric utilities. The assets are then allowed in what is called the utility's "rate base" on which public utility commissions set electric rates to provide a certain return (profit) to the utility. This caused the utilities to become profitable, stable, and credit worthy, often with excellent bond ratings. This credit worthiness would later become the financial basis of the non-utility generation (NUG) industry, later called the independent power producer (IPP) industry as we know it today.

3.5.2 PURPA and the Development of Non-Recourse Financing

The NUG industry was enabled by PURPA in 1978, but the first major financing was not until 1981 and it was an interesting case.

General Electric received a request for proposal from a major customer, Big Three Industries, a chemicals company in Houston, for an 80 MW gas-fired cogeneration facility under PURPA. That is, they were asking GE to design, build, and operate the facility, selling electricity to Houston Lighting & Power, and steam to Big Three Industries.

GE had never before done business this way but saw it as part of the future. The situation required GE Power Systems Sector to establish a sector-wide Marketing Council, chaired by Ed Phelps and managed by the author, to plan such a NUG offering. Those were interesting meetings as the bid required the gas turbine, steam turbine, generator, transformer and installation & service engineering divisions to collaborate—which really meant negotiating the split of profits from the project.

Making it even more complicated, the financing would need to be provided by GE Capital, which, until this case, was forbidden by a Justice Department decree in the 1950s preventing GE Capital from financing the sale of GE equipment because of the potential for "predatory pricing" by a market share leader. Until then, GE Capital had grown its business financing aircraft and railroad cars, principally by leasing. It was quite successful. GE at that time had a AAA credit rating, so it was borrowing at a AAA rating (low interest) and financing customers (at a higher BBB rating), hence arbitraging interest rates while also gaining the investment tax credits in the leases for the parent company. Very profitable!

After getting special approval by GE's Board of Directors with concurrence from the Justice Department for the GE Capital financing, GE retained the law firm Skadden Arps in New York City to prepare the legal documentation—the site acquisition, power sale agreement, steam sale agreement, equipment supply agreements, EPC (engineering, procurement, and construction) agreement, and financing documents.

On the first review, the project team laughed out loud when they read the loan documents—GE as project sponsor and owner would be liable to GE Capital as lender for repayment! No one had thought it through until that moment. So Skadden was sent back to rewrite the loan documents to say that this would be a "non-recourse" loan, for which the credit analysis would "look through" the power sale agreement to the credit worthiness of HL&P, and "look through" the steam sale agreement to the credit worthiness of Big Three Industries.

There was born the "non-recourse" financing—that is, not recourse to the project owner—that became the standard method of project financing that we use today for over $100 billion/year of solar financings.

Of course, Skadden had other clients coming into the new NUG business, and presented this financing solution to them. It quickly became the industry standard method of doing business.

3.5.3 Conditions Required for Project Financing

The first requirement of such large-scale project financing is a legal system that includes strong and stable contract law and financing law, and more specifically requires utilities to interconnect nonutility generators and purchase the power. The approach was born under US law, and has had difficulty being promulgated around the world, even in Europe when the IPP industry went there in the 1990s. What we have in the United States that is vitally important to the nonrecourse financing of power generation includes

- federal law in the form of PURPA, EpAct of 1992, the Energy Security Act of 2005, the Energy Act of 2007, the Economic Recovery Act of 2008, and the ARRA of 2009—a lengthy history of laws intended to enable low-cost financing of power generation;
- state laws regulating electric utilities, which includes approving Power Purchase Agreement (PPAs);
- a history of regulatory and court decisions under the legal system that supports the PPA approach and other aspects of being a nonutility generator;
- credit worthiness of the utility industry that is on the purchasing side of the PPA, as this will be the credit worthiness on which the project financing will be structured and priced;
- an experienced set of players in the project with excellent track records, including, but not limited to, the developer, equity investors, contractors such as the EPC contractor, the operations and maintenance (O&M) contractor, attorneys, consultants, and others.

Imagine the state of affairs today, as the solar industry goes global into developed and developing countries, seeking to develop projects with the above approach to project financing, where little or none of the pre-conditions exist. What we are witnessing is that only the massive developers are succeeding, such as ENEL, EdF, Iberdrola, E.ON, and others that have local presence and sufficient power in the marketplace to cause the host governments and utilities to honor their commitments with them over the lives of their projects, even if the laws do not provide clear protections. One of the leading figures in this field has been attorney Keith Martin at the law firm Chadbourne & Parke, who has advised many companies in the establishment of viable legal structures for projects.

3.5.4 Overall Capital Structure: Equity, Tax Equity, and Debt

The first aspect of developing a financing is structuring the "capital stack," which can include senior debt, subordinated or mezzanine debt, preferred equity (which typically today is in the form of tax equity), and common equity.

If tax equity is going to be used, then it is the first element to be determined, according to the rate of return required and the time period required to achieve that return (the richer the project, the shorter can be the achievement of the necessary tax equity return).

The next factor is deciding whether there will be traditional senior debt in the project that would be senior to the tax equity investor. It should be noted that many tax equity investors will not invest in a deal with senior debt above them—they want first call on the cash flows. If there is debt in the project, then the amount has to be determined according to standard practices, such as a minimum Debt Service Coverage Ratio (DSCR) over the life of the project. This is modeled in a spreadsheet analysis. Let's say, for example, that a lender wants to have a DSCR of 1.5× every year, which is to say that after-tax cash flows at 1.5 times the debt payment. This will determine how much debt the project can take on.

Then, there is the assessment of risks. There will be probabilities applied, such P50, which is the 50/50 likelihood, and a P90, which determines a 90% likelihood of happening, and finally the P99, which determines results to a 99% (near absolute) certainty. These are ways of quantifying the sense of risk as seen by experts with experience on such things. These figures are then used to structure and negotiate the debt portion of the financing.

Lastly, the amount of equity capital is determined—being the amount still to be supplied after tax equity and debt (if any) have been determined.

A simplistic view of typical project financing is 80/20 debt/equity, 70/30, 60/40, or even 50/50. It all depends on the willingness of lenders (or tax equity investor) to provide their capital considering the likely returns and risks on that happening.

Another aspect of a project financing is the value of the collateral, that is, the power plant. If the project stumbles financially, can the power plant be sold for enough to cover the whole investment, or just the debt (the equity investor is wiped out), or not even the debt (now a Chapter 11 or Chapter 7 bankruptcy situation). The tough one is the middle one, where the owner is willing to accept an outcome in which the lenders get repaid but equity gets little or nothing. But one thing is true, that no one has ever lost "all their money" on a power plant because there is an asset there that can be sold. This is the second part of technology risk. This first part is whether the plant will work as promised over its life and produce electricity as planned at the cost as planned. The second part is, if the project fails and the equipment has to be sold, whether it will sell for enough to repay lenders and investors? It appears that PV is proving to be low risk, while CSP may be higher risk, in this regard.

3.5.5 Tax Equity Using the Investment Tax Credit

The US Government has enacted a Production Tax Credit (PTC) for wind power and other kinds of renewable energy, and an Investment Tax Credit (ITC) for solar energy and several other technologies.

The solar ITC was first enacted in 2005, then extended in 2008 for an 8-year period to the end of 2016, then extended again at the end of 2015 on a scheduled ramp-down to zero in 2023. The schedule is as follows:

	Residential	Utility and C&I
2016	30%	30%
2017	30%	30%
2018	30%	30%
2019	30%	30%
2020	26%	26%
2021	22%	22%
2022	10%	0%
Thereafter	10%	0%

There are a number of complex rules to be followed in using the solar ITC, including, but not limited to, the basis being the cost of the system or fair market value, and the need to maintain ownership for a minimum of five years; otherwise there can be a "claw back" of the credit.

There are many key people running the tax equity space, but among them the leader is certainly John Eber at JP Morgan Chase in Chicago, who executes perhaps $3 billion of the total market $12 billion of tax equity invested each year.

3.5.6 Bank Loans

The major portion of capital provided to solar energy projects has been senior debt provided by banks. A typical deal will have a capital stack of 30–40% senior debt, 40% to 50% tax equity (initial "owner"), and 10% to 20% sponsor equity (long-term owner).

However, US banks are increasingly constrained from making loans with the needed tenors of 10, 15, and 20 years. This is different for European and Japanese banks, which come under a different set of risk-avoidance regulations. Indeed, there was a movement of bank capital from Europe to US renewable energy projects in the 2000s because such investments were deemed lower risk than putting the money to work in the EU. This also applied by Japanese banks that came into the US renewables market after 2010.

It is said by leading attorneys who do work for such project financings that there are over 100 banks competing for senior debt lending to US renewable energy projects. Similarly, it is said that there is plenty of debt capital for renewable energy projects in Europe.

In China, of course, the situation can be characterized as banks looking to place capital into renewable energy projects because they have been designated "priority" markets.

It is important to see that there are two worlds of renewable energy project financing: the upper 50% percentile of projects that the banks compete to lend to, and the lower 50% percentile of projects (and sponsors) that complain that there is not enough capital.

Several of the kingpins of this market include Sandip Sen and Marshal Salant at Citi, Ray Wood at Bank of America Merrill Lynch, Brian Bolster at Goldman Sachs, and others. The original Wall Street leader on renewable energy finance was John Cavalier at Credit Suisse, then at Hudson Clean Energy Partners, and now at Pegasus Capital.

3.5.7 Institutional Capital

Sources of longer-term debt for power plants include insurance companies, pension funds, and other investors seeking a stable, reliable, long-term return to match their long-term liabilities on insurance payouts and pensions.

However, insurance investors have technical difficulty using the ITC, and pension funds (not being tax-paying entities) cannot use the ITC at all. Thus, these logical investors in solar projects are if effect disadvantaged from participating in the financing of solar projects and installations. They do participate very well outside the United States. where the market is not driven by tax equity.

Several of the original leaders among insurance company investors included Herb Magid and John Buehler at Energy Investors Funds, and John Anderson at Hancock.

3.5.8 Project Bonds

The "project bond" market is smaller than many people believe, because there are so many known and unknown risks on investment anywhere outside the United States. This is relevant because, typically, project bonds are nonrecourse (no guaranteed or backstopped) by the sponsoring entity and are instead dependent on repayment from the cash flows of the project, which in turn depends on two things: that the project works and that the off-take purchaser will pay the project in full and on time, month after month.

Project bonds can only be issued for the highest quality solar projects that are developed by the highest quality sponsors. The dominant underwriter of project bonds in the world is Citigroup, where Stuart Murray has established a leading position in this area.

3.6 Capital Market Investment in Solar Securities

Michael Eckhart

The end game for 40 years of the advancement of solar technologies, industries, markets and policies has always been envisioned as the entry into the capital markets—the massive markets for stocks (equity) and bonds (debt). This is where investors compete for the least capital at the greatest returns and issuers compete for the most capital at the lowest cost, and investment bankers bring the two together, generally representing the issuers but always with an eye to meeting the needs of the investors.

3.6.1 Equity Market Investment in Solar Companies

The best understood capital market is the equity capital market, generally called the stock market by most people.

Solar PV companies started emerging as financially attractive enterprises since about 1995.

The original solar PV startup was Solarex, co-founded in 1973 By Joseph Lindmayer and Peter Varadi. Solarex was sold in 1983 to Amoco Oil and operated as a wholly owned subsidiary until Amoco was merged into BP in 1998.

In the mid-1990s, the Japanese PV companies emerged but they were not start-ups looking for their initial public offering (IPO). They were companies that were already publicly traded like Sharp, Kyocera, Fuji and others.

Then in the late 1990s came the PV divisions of large European companies like Siemens Solar, Shell Solar, and BP Solar, all being publicly traded already.

Then in the early 2000s things began to change as new start-ups began emerging as viable PV suppliers and went public, such as

- First Solar: previously named Solar Cells, Inc. founded in 1990, reborn as First Solar by CEO Michael Ahearn in 1999, and IPO in 2007 on the Nasdaq market;
- Q-Cells: founded 1998 by CEO Alton Milner and IPO in 2005 on the Frankfurt stock market;

- SunPower: founded 1999 by Stanford Professor Richard Swanson and IPO in 2005 on the NASDAQ market.

These companies were the first to "reach the stock market" heralding the real success of the PV industry.

Then came the great PV revolution by China that kicked off with the passage of China's Renewable Energy Law in 2006.

There was a magical match-up between Europe's Renewable Energy Directive in 2004—that called for Europe to adopt renewables, create a renewable technology manufacturing industry, and be the world leader in exporting renewables (and renewable technology) to the world—just as China's renewable energy law called for the development of a solar PV manufacturing industry with an export focus. To succeed, China needed three things:

- cheap capital to fund the factories, supplied by the Chinese banks
- PV manufacturing technology, which ironically Europe was actually subsidizing
- international market demand, which, again, Europe was ready to supply

The key was the European subsidies of European PV factory technology companies. Astonishingly, European technology suppliers were being subsidized to provide factory technology to China, whose factories would export back to Europe and put the European PV industry out of business.

The Chinese were absolutely brilliant about this, using the free capital from Chinese banks to buy European technology and sell the PV cells and modules back to Europe. It was an amazing combination of circumstances.

This led to immediate success for Chinese PV companies. By 2010, there were already 523 PV manufacturing companies in China according to the China General Certification Center, a division of government that certified all manufacturers. Over a dozen Chinese PV companies went public in a variety of markets (Yingli, Trina, JA, Canadian Solar, GCL, Jinko, others). One of the great Chinese entrepreneurial stories is about Jifan Gao, founder and CEO of Trina Solar which went on the be the world leading PV manufacturer in 2016.

The publicly traded PV manufacturers have struggled to maintain support by public stock market investors. The massive oversupply of PV modules from China created competitive conditions in the 2010–2015 period such that prices were driven down and costs had to be eliminated as fast as possible, but not fast enough and profits were driven out of the industry. Indeed, this was an historic shift of profits in the PV business overall from PV manufacturers to PV project developers and installers as the market price of modules came down rapidly.

This shift led to the success of PV developers such as SunEdison in the United States and Conergy in Europe, both of which became publically traded companies—Conergy by going public in Frankfurt and SunEdison by being acquired by the publically traded raw silicon producer MEMC, which changed the whole company's name to SunEdison.

The shift of profitability from manufacturing to installations also is supporting the rapid growth of residential rooftop installers and financiers such as SolarCity, SunRun, and Vivint that have done IPOs. There are others emerging, too, but are still privately held as of this writing. Further, there are financing specialists such as Spruce Capital (formed by the merger of Clean Power Finance and KW Financial), Renew Finance, Mosaic, and Admirals Bank, among others.

So by 2015–2016 the solar PV industry had reached the stock markets and raised a substantial amount of capital but was still struggling to make a profit in manufacturing and persuade investors that there is a viable "business model" in the development business (Abengoa and SunEdison have filed Chapter 11 bankruptcy) and in the "resi-rooftop" business.

The PV industry continues to experience booming growth in the markets while individual companies crowd the field competing for a slice of the business. One only hopes that the PV installation industry will mature and stabilize (and keep growing) in the coming years.

3.6.2 Yieldcos and Other Portfolio Companies and Funds

An important development occurred in 2013, in which a new model of renewable energy company was created called a "Yieldco."

Here is how it happened. After 2010, when wind and PV markets were taking off, there was some frustration in the task of raising capital for the projects at the lowest possible cost of capital especially the cost of debt capital.

Other industries had already successfully lobbied the US Congress to allow special financing vehicles for portfolios of operating assets that were deemed to be there for efficiency and not for making a profit per se, so they are not taxed at that level. There are two types. First is the Real Estate Investment Trust (REIT) for real estate projects and assets. Second is the Master Limited Partnership (MLP) that was created as a non-tax paying entity to hold the operating assets of oil and gas pipelines and other energy and forestry projects.

The renewable energy industries have been lobbying for some time that they should be allowed such a vehicle for ownership of operating assets like wind farms and solar projects. This is still under consideration by the US Congress.

One company that wanted to get this treatment was NRG Energy, headed by CEO David Crane. In preparation for the Renewable Energy MLP passing Congress, Crane asked Citigroup to do an analysis of NRG's portfolio of renewable energy projects and recommend which of them would get put into the MLP. An analyst stayed late into the night doing the analysis and came to the meeting in the morning with two recommendations instead of one. He said, these would be going into the MLP, but these others will not be paying taxes for more than 10 years and NRG could spin them off into a taxable "Inc." corporation that would work effectively, as an MLP. This was then presented to Crane and his team, and Crane, being an ex-Lehmann Brothers investment banker himself, approved and said they would call it NRG Yield, Inc. The Citi bankers tried to argue back, saying that such a name undershot the full potential of the new companies because it would have current yield PLUS managed growth, giving a "total return" much higher. Say, for example, the current yield from dividends might be 6% plus growth of 10%/year, then providing a total return of 15%. The Citi bankers suggested calling it a "total return co" but Crane said: "No, I like "yieldco" and that was that. The Yieldco was thus invented. NRG Yield did its IPO on July 22, 2013.

Then in rapid succession, the bankers persuaded clients to implement the new Yieldco model, including

- Pattern Energy Group (PEGI) September 2013
- Abengoa (ABY) June 2014
- NextEra (NEP) June 2014
- SunEdison created two yieldcos: TerraForm Power (US assets)—July 2014 and Terraform International (global assets) 2015
- First Solar and SunEdison created a jointly owned yieldco named "8Point3" June 2015

Things seemed to be going well. Sophisticated investors experienced in the power business were buying the yieldco IPO shares. But then for reasons now totally understood, the air went out of the balloon almost immediately after TerraForm International went public. Perhaps investors realized that there was still risk in these assets. Perhaps they realized that the promised growth was only made possible by the yieldcos raising more capital from them—since almost all profits were reserved for issuance of dividends to investors, there was little or no internally generated capital for growth. Then yieldco share prices took a nosedive starting in July 2015 losing an average of 50% of their value in less than six months.

President Dwight D. Eisenhower signed the REIT law in 1960. REITs are corporations that own and operate real estate properties and distribute their income to the shareholders. The REIT idea became so popular that at present REITs exist in 37 countries and in 2014 were listed globally on 456 stock exchanges. The other is called a Master Limited Partnership (MLP) that is used for natural resource related assets like oilfield equipment and gas pipelines.

The only yieldco that has gone public as a REIT is Hannon Armstrong Sustainable Infrastructure Capital, Inc., which went public in April 2013 and trades as HASI, headed by CEO Jeffery Eckel. HASI was a unique company, having created a market for securitizing Federal Energy Savings Performance Contracts (ESPC), of which HASI completed over $4 billion of such securitizations, selling the securities to insurance company investors. Because energy efficiency assets are deemed to be "REITable," this gave HASI a substantial platform for going public as a REIT. Another relevant rule is that a REIT must have no less than 75% of its assets as REITable assets; this gave a window for investing as much as $1 billion into renewable energy (or other) assets. No other company has HASI's EE asset base, so none can duplicate this unique financing achievement.

3.6.3 Green Bonds

The first green bond was issued by the World Bank in 2008. This was accomplished by a banker named Christopher Flensburg at SEB in Sweden, who had an investor client that wanted to buy a bond that had an environmental use of funds. Having an investor ready to buy, Flensburg went in search of an issuer. He found a receptive audience at the World Bank, which issued a one-investor "green bond." That was the beginning of a massive upwelling of the issuance of green bonds in years later.

There is some friendly competition for being credited "first" with the European Investment Bank (EIB), which issued an "impact bond" in 2007, saying that this was the first of the environmental purpose bonds. So credit both the World Bank and EIB for getting things started.

The definition of a green bond emerged over the next several years as the multilateral development banks (MDBs) issued green bonds from time to time, and continue to do so. Then there were several green bonds issued by municipalities and others.

Things began to brew in 2012–2013, during which time there was a spreading interest and discussion about "climate bonds" and "sustainability bonds" and "green bonds." There was growing interest among potential corporate issuers, municipal issuers, government issuers, and others. There was a great debate about what is a green bond, and what is it not; about who can issue one and who cannot; and about what the rules and penalties for breaking the rules should be.

In February 2013, the International Finance Corporation (IFC, a unit of the World Bank Group) issued a $1 billion green bond. Citigroup was one of the placement agents. It was a tremendously successful placement, with sale of the bonds reaching the $1 billion level in one hour. This was astonishing, as a normal successful placement takes one to two days in order to be able to call investors in all parts of the world. Citigroup did an informal review of how this success happened—what was IFC doing that caused such a rush to invest? This would become useful several months later.

Three months later in April 2013, a publishing firm named Environmental Finance held their annual conference in London, where the debate about green bonds came to a head. An element of the discussion was about how there was a void—a complete lack of any written guideline of any kind on the issuance of green bonds.

At the conference and following, there was a conversation among Michael Eckhart from Citigroup, Susan Buchta from Bank of America Merrill Lynch, and Sean Kidney from the Climate Bonds Initiative (CBI), on the need for a guideline to protect the integrity of a green bond market, to prevent any "green washing" or other mistakes and misbehaviors. Kidney's approach was to develop a better definition of what qualifies for a green bond, more or less defining what is green and declaring a standard around that definition. Eckhart and Buchta said that's fine, but we really need a process guideline—that is, guidance on what an issuer must do to qualify its bond as a green bond, as to its honesty and integrity, to ensure that there will be no cheating or misrepresentations about green bonds.

Through the summer months of 2013, Eckhart and Buchta drafted a document called the Framework for Green Bonds that was renamed the Green Bond Principles (GBP) in September when the working group was expanded by adding bankers from J.P. Morgan and Credit Agricole. The development work rolled forward to January 2014 when the GBP was put into force by the adoption by 13 banks from the United States and Europe. A system of Governance was added in April 2014, along with appointing the International Capital Market Association (ICMA) as Secretariat of the Governance. At that point, the "guideline" that was under debate a year earlier had been put in place.

The GBP is a voluntary, self-regulatory process guideline for issuing a green bond. Its purpose is to cause green bond issuers to disclose certain information to prospective bond investors so that they can make informed decisions knowing the use of funds. It has four requirements, that an issue:

- first, declare its investment criteria for the use of funds;

- second, define its investment decision making process;
- third, keep the use of proceeds sequestered or accounted for separately to ensure that the funds are only be used for green purposes even if temporarily;
- fourth, agree to report after the fact on what specific projects or purposes the funds were used for, so that investors can calculate the environmental or climate impacts of their investment in the bond.

Interestingly, the GBP is the first use of funds guideline in the 200-year history of the bond market, where, traditionally, the statement on use of funds has been "general corporate purposes" or equivalent words.

The issuance of green bonds increased immediately after promulgation of the GBP, from $13 billion in 2013 to $36 billion in 2014 and $42 billion in 2015.

3.6.4 Securitization

Asset-backed securities (ABS) are a class of bonds for which the source of payment of principal and interest is tied specifically to the cash flows from an underlying set of assets such as solar loans and leases.

The ABS class is much broader that solar loans and leases including, for example, bonds based on payments from auto loans, student loans, credit card receipts, and other assets.

The ABS class is modeled after the well-known mortgage-backed securities market that is promoted by Fannie Mae, Freddie Mac, Ginny Mae, and private investment banks.

Solar ABS is the new kid on the block, having the first "solar securitization" completed by SolarCity, placed by Credit Suisse, in July 2013 (interestingly enough, the same month in which the first Yieldco was created). Indeed, we might look back to see 2013 as the inflection point in the innovation of financing for clean energy.

The keys to success in the ABS business includes large scale (typically 10,000 or more underlying financings), uniformity of contracts (so lawyers don't have to read 10,000 documents), robustness, depth and longevity of the database evidencing the creditworthiness of the underlying financings, and history of similar transactions. That is to say, the elements of being smaller scale, heterogeneous, lacking a credit database and new—which characterizes the status of solar securitizations in 2016—makes it challenging to get solar securitizations going.

The process for preparing and issuing a solar or other ABS securitization illustrate why it is difficult to get the ball rolling. There are 10 basic steps:

(1) making the original financings

(2) issuing a mandate to an investment bank to underwrite and manage the placement of the asset-backed securities

(3) having the investment bank create a "warehouse" that will acquire the loans or leases and "store" them for a period (typically six months to a year) to prepare legally everything for the issuance of the bonds

(4) seeking a credit rating for the issuance from Standard & Poor's, Moody's, Fitch Ratings, Kroll Bond Ratings, or other credit rating firm

(5) preparing the legal documentations for the issuance

(6) preparing and filing documents with the Securities and Exchange Commission (SEC) and any applicable state regulatory agencies

(7) conducting pre-marketing of the issuance

(8) pricing the issue

(9) obtaining final commitments from investors

(10) completing the issuance and securities distributions to investors

Successful securitizations have been completed by SolarCity, SunRun, and Renew America, and additional securitizations are being planned by Spruce Capital, Renew Financial, Mosaic, and others.

3.7 Summary

Michael Eckhart

The financing of solar energy has traveled an interesting journey since the late John Bonda invited this author to join a renewable energy trade mission, sponsored by the European Commission, to South Africa in late 1995. He could not locate anyone in Europe to give a talk on solar financing. So here he was inviting an American to join a European trade mission. That most likely defined the starting point for solar financing, where, in South Africa, I presented the case for shifting the industry's thinking form $/watt to cents/kWh by financing the installations.

Today, solar financing is a $100+ billion/year business and it seems headed to be 10 times that level, perhaps someday $1 trillion/year being invested in solar energy.

We are on a tear. May it continue forever until solar energy is providing 100% of the energy needed to support humanity and the rest of our world in a sustainable way.

3.8 Glossary (For Chapters 3.3 to 3.7)

ACORE : American Council On Renewable Energy

ADB : Asian Development Bank

ARRA : American Recovery and Reinvestment Act of 2009

CREIA : European Renewable Energy Council

CSP : Concentrated Solar Power

DOD : Department of Defense

DOE : Department of Energy

DSCR : Debt Service Coverage Ratio

EBRD : European Bank for Reconstruction and Development

EIB : European Investment Bank

EPC : Engineering, Procurement and Construction

EPRI : Electric Power Research Institute

ERDA : Energy Research & Development Administration

EREC : European Renewable Energy Council

ESPC : Energy Savings Performance Contract

FERC : Federal Energy Regulatory Agency

FPA : Federal Power Act

IEA : International Energy Agency

IFC : International Finance Corporation

IPP : Independent Power Producer

IRENA : International Renewable Energy Agency

ITC : Investment Tax Credit

JPL : Jet Propulsion Laboratory

MUSH : Municipals, Universities, Schools and Hospitals

NASA : National Aeronautical and Space Administration

NEM : Net Energy Metering

NUG : Non-Utility Generation

NYSERDA : New York Energy Research & Development Agency

PPA : Power Purchase Agreement

PSA	:	Power Sales Agreement
PPTC	:	Power Production Tax Credit
PTC	:	Production Tax Credit
PURPA	:	Public Utility Regulatory Policies Act
REC	:	Renewable Energy Credit
REIT	:	Real Estate Investment Trust
REN 21	:	Renewable Energy Network for the 21st Century
RPS	:	Renewable Portfolio Standards
SERI	:	Solar Energy Research Institute
SPA	:	Special Purpose Vehicle
SREC	:	Solar Renewable Energy Credit
UNFCCC	:	UN Framework Convention on Climate Change
WCRE	:	World Council for Renewable Energy

4

Present and Future PV Markets

4.1 The Global View of PV

One of the previous sections described the new markets that not only sustained the mass production of PV but also drastically increased the volume of production resulting in the substantially decreased cost of PV systems. As a result, the price of the electricity produced by PV systems became highly competitive with the price of electricity generated by any other means. This increased again the demand for PV modules.

Another previous section described the development of the financial mechanism to provide more and more money needed for financing this increased volume of PV systems.

This section provides the exact information and statistics how the demand influenced mass production of PV and the development of financing influenced the PV markets in the Americas, Australia, Asia, and Europe and also in the neglected PV markets of Africa and the Middle East.

Finally, renewable energy, of which PV is the flag bearer, was accepted as one of the major energy production systems. This happened many years after the nuclear and fossil energy received global recognition.

The International Atomic Energy Agency (IAEA) was initiated in 1957 to promote the peaceful use of *nuclear energy*.

The International Energy Agency (IEA) was founded in 1974 after the 1973 oil embargo in the framework of the *Organization for Economic Co-operation and Development* (OECD). The IEA was initially dedicated to respond to physical disruptions in the supply of *oil* and also served as an information source on statistics about the international oil market and other *energy* sectors. It was stoutly opposed to renewables. The IEA routinely neglected the position of renewables and always underestimated the growth potential of solar and wind.

PV Industries were only represented by the various countries' or regions' industry associations. The PV industry associations had poor financial support from the fledgling PV industry and

Sun towards High Noon: Solar Power Transforming Our Energy Future
Peter F. Varadi
Copyright © 2017 Peter F. Varadi
ISBN 978-981-4774-17-8 (Paperback), 978-1-315-19657-2 (eBook)
www.panstanford.com

none from governments. They operated independently and were not even connected to one another. The idea of an International Renewable Energy Agency (IRENA) came up around 1990, but was only established in 2009 as an UN organization supported by many countries.

IRENA is described in this section for two reasons. It is an important organization for PV. It is important because it is connecting the utilization and the industries of PV to all member countries of United Nations. It is also important because it is collecting relevant information even about the number of jobs in renewable energy on a global scale and providing valid information for governments, for financial institutions and marketing.

4.2 The Present and Future of Neglected PV Markets: Africa and the Middle East

Frank P. H. Wouters[1]

4.2.1 Introduction

As a young engineer, in the early 1990s I worked in Zambia, a beautiful landlocked country north of Zimbabwe. Although Zambia has a clean energy system, which relies almost entirely on hydropower, the problem in Zambia was and still is that the electricity grid does not reach very far into the countryside. So most people use diesel pumps and diesel generators for farm and household applications, and like in most rural areas in Africa, most people don't have electricity at all.

I was head of the design section at a business unit connected to the School of Engineering at the University of Zambia. Our aim was to develop appropriate technologies and introduce them in the local market. We closely cooperated with a Jesuit mission post in a rural area outside of the capital city Lusaka. They had problems with unreliable grid electricity and also operated many diesel generators on remote locations, for example to pump water. The diesel generators were expensive to operate and failed often, requiring expensive spare parts and repairs. The supply of diesel was not very reliable, so very often there was no water.

One of my first projects with them was the design of a wind pump. The well-known farm wind pumps that can be seen throughout the United States were also present in Africa. However, they had a rather complicated gearbox, which could not easily be manufactured locally, so we used a special direct drive design, whereby the pump rod was directly connected to the crank. The short crank was connected to the main shaft, which was propelled by the rotor. Our rotor had to run faster than with a gearbox, so the design had to be modified for that. In fact, I went into town and bought a standard valve pump used in hand pumps, and modified it to fit our design. After we had manufactured the components, we transported everything to the Jesuit mission post outside of Lusaka. We had a borehole drilled with water level some 26 meters below ground. We assembled the wind

---

[1]Frank P. H. Wouters' biography is on page 294.

pump, which was more than 10 meters tall, lying on the ground, and used a car winch to hoist it up. We lowered the pump deep enough in the water and connected everything. We installed the water tank next to the wind pump. Then we waited for wind and were excited when the wind pump started running and the tank started filling up. Within a few hours we had filled the tank, and the water could be used for high-value crops and for cattle drinking water at a location where that was not available before. Next, I took our design to a local company that supplied diesel pumps to local farms. They knew there would be a market for a product that doesn't need electricity from the grid or expensive diesel. At the time the cost of solar panels was still rather high, so solar PV pumping systems were not widely seen in Zambia. They really liked the simplicity of the wind pump we designed, since it could be manufactured and repaired locally, instead of importing components from abroad. We made some jigs for them so they could start series production and they successfully introduced our design into the Zambian market.

This experience taught me a few lessons. First, using renewable energy makes a lot of sense in places where it is difficult and expensive to buy diesel or petrol and carry out maintenance on complicated machinery. Second, although the efficiency of the equipment is important, reliability and robustness are more important in harsh environments. The slow-running farm wind pump, which has a gearbox, produced more water than our simpler design. However, the gearbox could not be sourced locally and if there is a problem, replacement takes very long, requires foreign currency and might lead to the failure of the entire wind pump. The same principle is valid for solar equipment. Very often I see designs made by Western educated engineers that are highly efficient but not very robust. I have seen sophisticated battery management electronics, for example, or advanced pump controllers that work fine in a moderate climate and sophisticated environment where skilled engineers and spare parts are available. However, when you are in a rural area in a developing country, several hundred kilometers away from the first major town, it is more important to keep it simple and make something that works and can be fixed locally, especially in a hot and dusty or humid environment, with aggressive insects, hostile to electronic equipment.

4.2.2 Africa

After nightfall, most of the African continent is still in the dark, more than a century after the light bulb was invented. According to the World Bank, some 25 countries in sub-Saharan Africa are facing an electricity crisis evidenced by rolling blackouts. On average, the blackouts cost more than 2% of GDP, not factoring in the potential for additional economic activity if the supply of generation capacity would match real demand. But worse still, 76% of sub-Saharan Africans, 620 million people in total, are not affected by the blackouts, since they don't have access to electricity at all.

The comparison with South Asia, which has similar per capita incomes, is particularly striking. In 1970, sub-Saharan Africa had almost three times the generating capacity per capita as South Asia. In 2000, South Asia had left sub-Saharan Africa far behind—with almost twice the generation capacity per capita.[2]

However, Africa, which in the 1980s and 1990s used to be labeled the "lost continent" by many, has been cast in global spotlight thanks to its impressive economic growth since 2000, leading to what people now call the "African century." The majority of the world's fastest growing economies since then have been from Africa and many countries are finally on the way to become middle-income countries, with extreme poverty declining and the middle class growing. However, the necessary infrastructure for sustained economic growth is inadequate and its development is lagging. Also, the total number of people without access to modern energy is rising because electrification efforts are outpaced by population growth.

For an informed discussion, one has to differentiate between sub-Saharan Africa and the countries in the Middle East and North Africa, the MENA countries that are bordering the Mediterranean Sea and the Arabian Gulf. In the MENA region, more than 99% of the population has access to electricity, whereby more than half of sub-Saharan Africans lack access.

Electricity in sub-Saharan Africa

Only Cameroon, Côte d'Ivoire, Gabon, Ghana, Namibia, Senegal, and South Africa have electricity access rates exceeding 50%.

[2]Vivien Foster and Cecilia Briceño-Garmendia, eds. (2010). *Africa's Infrastructure: A Time for Transformation*, AFD/WB, Washington, D.C.

The rest has an average electrification rate of 20%. According to McKinsey,[3] it takes 25 years to progress from a 20% electrification rate to 80% electrification rate. Furthermore, there is a strong link between electrification and GDP per capita, typically with steep growth once a country reaches access rates above 80%. It is therefore now time for decisive action. Although the lack of power holds back growth, African resourcefulness always finds solutions, albeit at huge cost. Many people and businesses have their own generators, and it is estimated that in Kenya more than half of businesses own generators, while the estimated number of generators in Nigeria is a staggering 60 million. The cost of electricity from such diesel and petrol generators is on average four times the cost of grid electricity. In Nigeria, diesel fuel is a leading expense for the major African mobile-phone companies, representing up to 60% of operators' network costs.[4] African enterprises have identified unreliable power supply as the most pressing obstacle to the growth of their businesses, ahead of access to finance, red tape, or corruption.[5]

The installed capacity in sub-Saharan Africa in 2012 was 90 GW, half of which is in South Africa. Forty-five percent of this capacity is coal (mainly South Africa), 22% hydro, 17% oil (both more evenly spread), and 14% gas (mainly Nigeria). Affordability is a critical issue, since electricity prices are typically very high by world standards, despite often being subsidized. Electricity consumption per capita is, on average, less than 1,000 kWh per year, which is less than that needed to power two 50 W light bulbs continuously.

Table 4.1 shows the installed solar PV capacity of African countries at the end of 2015 as reported by IRENA. The capacity includes grid-connected as well as off-grid applications and is based on the overall data on import of solar modules, which is a pretty accurate measure of what has entered the countries. Although it is very difficult to extract exact data on off-grid applications, if we subtract a reported 1,362 MW of operating

---

[3]Antonio Castellano, et al., McKinsey and Company (2015). *Brighter Africa: The Growth Potential of the Sub-Saharan Electricity Sector.*
[4]Emmanuel Okwuke (February 2014). Nigerian telcos spend N10b yearly on diesel to power base stations—Airtel boss, *dailyindependentnig.com.*
[5]Fatih Birol, et al., IEA (2014), *Africa Energy Outlook: A Focus on Energy Prospects in Sub-Saharan Africa.*

utility-scale plants with a capacity larger than 100 kW as reported by Platts,[6] the off-grid solar PV accounts for roughly 35% of the overall volume. Angola, Comoros, Cote d'Ivoire, Democratic Republic of the Congo, Equatorial Guinea, Eritrea, Ethiopia, Gabon, Guinea, Liberia, Sao Tome and Principe, Sierra Leone, Sudan, and Swaziland reportedly had no solar PV capacity at the end of 2015.

Table 4.1 Installed solar PV capacity in African countries at the end of 2015

Country	MW PV	Country	MW PV
South Africa	1361	Mali	6
Algeria	273.6	Niger	6
Reunion	180.4	Libya	5
Egypt	25	Madagascar	5
Kenya	24	Zimbabwe	4.4
Namibia	20.5	Burundi	2.2
Uganda	20	Ghana	2
Mauritius	18.2	Zambia	2
Mauritania	18	Botswana	1.7
Morocco	17.8	Cameroon	1.5
Nigeria	17	Benin	1.3
Tunisia	15	Malawi	0.9
Tanzania	14	Somalia	0.6
Mayotte	13.1	Togo	0.6
Cabo Verde	10	Congo	0.5
Rwanda	8.8	Central African Republic	0.3
Senegal	8	Djibouti	0.3
Burkina Faso	7	Lesotho	0.2
Mozambique	7	South Sudan	0.2
		Seychelles	0.1
		OVERALL	**2,099**

Source: IRENA.

---

[6]www.platts.com.

Table 4.1 shows that at the end of 2015, only one third of all African countries had more than 10 MW of solar PV capacity and only Réunion, Algeria, and South Africa had more than 100 MW. One has to consider that Réunion, a small island east of Madagascar, is one of the overseas departments of France. It is also an outermost region of the European Union and, as an overseas department of France, part of the Eurozone. It therefore has the same favorable regulatory framework for Solar PV as France, hence the comparatively large PV capacity.

Where the Grid doesn't Reach

In sub-Saharan Africa, 620 million people are not connected to the grid, and this number is rising despite spectacular economic growth in many countries. The number of people living off the grid has risen by 114 million on the continent since 2000, with several more million joining every year. Although many governments are working on expanding their electricity grids, many people live in sparsely populated areas far away from main transmission corridors. For such areas, extending the grid is a very expensive option, costing between $500 and $1,000 per connection, and the expected flow of electricity will be limited for years to come. Dedicated off-grid solutions are in many cases a more economical and appropriate solution. The standard solution so far was a diesel generator. The recent cost competitiveness of especially solar PV has made business models based on renewables much more compelling. Renewables don't need fuel and the lack of moving parts, especially in the case of solar PV, reduces the need for maintenance. A number of solutions using modern renewables to provide off-grid electricity have been around for some time. They include solar products such as solar lanterns, solar home systems, and mini-grids, but also dedicated solutions to power a business.

Solar Products

People who do not have access to grid electricity will serve their energy needs by other, often very expensive, means. According to a new report released early 2016 by the World Bank Group and Bloomberg New Energy Finance, in collaboration with the Global Off-Grid Lighting Association,[7] globally 240 million mobile

[7]https://www.lightingafrica.org/launch-of-2016-world-bank-group-bloomberg-off-grid-solar-market-trends-report/.

phone subscribers live off-grid. These phones are often charged by small businesses costing on average $0.20 per charge. Expressed in electricity, this equates to an astounding $30–$50 per kWh. For a proper cost assessment, one would have to include the travel time to the charging point, which can be substantial. This shows the enormous potential for cost-effective off-grid electricity solutions, including solar. Figure 4.1 shows that people spend $2.4 billion on mobile phone charging in Africa each year. But they also spend over $14 billion on kerosene, batteries, and candles for lighting, an amount almost as high as the GDP of a country like Zambia. The annual expenditure bill for household lighting varies from country to country but ranges $100–$140/year. Kerosene lanterns, a century old technology, are not only costly but also fire hazards. The wicks smoke, the glass cracks, and the light is often too weak to read by. The World Health Organization says the fine particles in kerosene fumes cause chronic pulmonary disease, affecting in particular women and children.

However, for a number of years, with solar panels becoming more and more affordable and LED lights becoming more and more powerful, so-called solar lanterns have started being sold in millions. A solar lantern consists of an electric lamp with a recharge-able battery that is charged with a small solar panel. If the solar panel is smaller than 10 W, the system is called a pico-solar system.

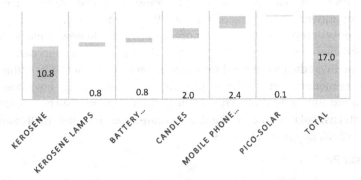

Figure 4.1 Estimated annual spend on off-grid lighting and phone charging in Africa (2014, $ billion; *source*: footnote[8]).

Figure 4.2 shows a solar lantern with phone-charging capability. Even the most basic solar lamps outperform kerosene

[8]Fatih Birol, et al., IEA (2014). *Africa Energy Outlook: A Focus on Energy Prospects in Sub-Saharan Africa.*

lanterns in terms of ease of use and quality of light. A typical solar lantern takes 8 to 10 hours to charge and then provides 4 or 5 hours of light from high-efficiency white LEDs. The number of times solar lamps can be charged before their rechargeable batteries wear out has improved enormously in recent years, along with their ability to cope with dust, water, and being dropped. A simple solar lantern can cost as little as $5, with higher quality models, for example with phone-charging capability, costing up to $50. Although this is not cheap for a poor rural household, the savings on kerosene and mobile phone charging provide for really compelling economics. A "typical" solar lamp costing $13 has a payback time of just a few months in most sub-Saharan African countries. Assuming a life of 2 to 3 years for a product from a reputable manufacturer, customers can enjoy at least one year of free lighting until they have to make another investment.

Until the end of 2015, more than 20 million units have been sold in Africa, with half of those supplied by reputable companies with a quality product, the other half, lower cost generic products. It is no surprise that quality products, requiring a price premium, have been copied by manufacturers of generic products, with lower quality components and a consequential shorter life or inferior service.

Figure 4.2   A solar lantern with phone-charging capability.[9]

Lighting Global[10] is the World Bank Group's platform supporting sustainable growth of the off-grid solar market. Through Lighting Global, International Finance Corporation (IFC) and the World

---

[9]http://sunnymoney.org/index.php/solarlights/.
[10]https://www.lightingglobal.org/.

Bank work with the Global Off-Grid Lighting Association (GOGLA), manufacturers, distributors and other development partners and end-users to develop the off-grid lighting market. An important aspect is the quality of the products, Lighting Global provides a quality assurance methodology with the following key aspects:

- **Truth in Advertising:** Advertising and marketing materials accurately reflect tested product performance.
- **Durability:** The product is appropriately protected from water exposure and physical ingress and survives being dropped.
- **System Quality:** The product passes a visual wiring and assembly inspection.
- **Lumen Maintenance:** The product maintains at least 85% of initial light output after 2,000 hours of operation.
- **Warranty:** A one-year (or longer) retail warranty is available.

Quality is important, since a survey done by SolarAid in Tanzania,[11] where a lot of low-quality products are being sold, showed customer satisfaction slipping from 97% for quality-verified brands to 60% for other brands due to lower product quality.

Solar Home Systems

One step up from pico-solar are solar home systems. In 1992 Anil Cabraal of the World Bank ASTAE division came up with the idea, which was named "Solar Home System" (SHS). These systems were believed to be a potential solution for the 1.6 billion people that do not have grid electricity. Solar home systems are individual household systems consisting of a solar panel, a battery, and some appliances.

Compared to conventional energy available in rural areas such as kerosene and disposable batteries, the cost of a solar home system comes up front. This typically requires some sort of financing, since most people in rural areas of developing countries cannot afford such a large amount. Although most developing countries have systems of rural credit supply, most traditional loans are given towards productive uses, for example, to buy seeds for the next year's crop, ensuring the bank there will be a future

---

[11]SolarAid (June 2015). *Research Findings: Baseline and follow-up market research in Kenya, Tanzania and Zambia.*

income to pay back the loan. Although clean household energy for light, mobile phone charging and radio/TV supports development in the long run, it is difficult for a bank to identify additional income in the time they require to recoup the money they lent to the rural customer. So solar home systems have traditionally been supported by development finance institutions (DFIs) such as the World Bank, the African Development Bank, and others, which supported local banks that had the infrastructure in place to provide such loans in rural areas. Also, 20 years ago one needed big solar panels and heavy car batteries to power inefficient appliances for an individual household. But now, with LED lights and efficient DC appliances, and low-cost reliable storage based on Li-ion batteries, which are much lighter and last longer than lead acid batteries, the overall energy demand is reduced. So a much smaller panel is required. And since the price of solar has fallen dramatically, the overall cost of the system is much more affordable now than 10–20 years ago.

M-Kopa

But it wasn't until another innovation became available, mobile money, that we saw a real breakthrough in the market uptake of solar home systems in East Africa. M-Kopa[12] is the company that pioneered the application of mobile money in combination with solar systems, first in Kenya and now in more and more countries. M-Kopa basically provides a financing solution so that people can afford solar PV.

In 2006, Chad Larson and Jesse Moore were doing an MBA at Oxford University when they met Nick Hughes. Nick presented M-Pesa, the first successful mobile money system as pioneered by him, when he was working for Vodafone in Kenya. The success of M-Pesa[13] intrigued them and they kept in touch with Nick in the years following his presentation. In the years between 2007 and 2010 they all had other jobs. In 2009, however, both Jesse and Nick left their jobs to explore business opportunities based on mobile banking. Both of them had a background in telephony and they were happy that Chad joined them in 2010, since he was a banker. In 2010, they successfully applied for grant money to

---

[12]http://www.m-kopa.com/.
[13]https://www.worldremit.com/en/m-pesa-mobile-wallets?gclid=CO3Tu8qdxs0CF UFehgodNFUMkA&ef_id=Ue2rSQAAAY8sp3yU:20160626173650:s.

pilot a few businesses based on mobile phones in Africa, among others from the Shell Foundation. They introduced a mobile savings account, started a company providing health services by linking into a network of medical doctors, and piloted a business providing solar home systems that incorporated a switch that can be operated remotely using a sim card in a modem. Of the three ventures, the solar venture was by far the most successful. They installed 300 pilot systems and 95% of the people actually paid back their dues because of that switch. The additional response from neighbors, etc., was overwhelming. They had so many requests for additional systems that they realized that this was a winning formula.

Their system works as follows: The client can charge credit on to the modem using his cell phone. If the credit on the solar system is finished, the system automatically switches off. The entire system is automated, customers get an SMS warning them of low credit, and they will be warned when the system is finally remotely switched off. The credit used is the normal mobile phone credit that can be bought at roadside stalls throughout the country.

At present people spend considerable amounts of money on kerosene and batteries. M-Kopa targets rural customers that typically have one or two phones in the household. Most households spend half a dollar each day on kerosene and batteries, and M-Kopa charges that amount for a much better solar electricity service. With new technology, solar can now easily replace kerosene and batteries, and supply enough energy for a DC radio, a flashlight and even a TV. The technology as marketed by M-Kopa actually provides the experience of having grid electricity. The system is supplied with cabling and light switches, so when people enter their house they can switch on the light immediately. A 20 W solar panel provides enough electricity to charge a Lithium-ion battery, with a 5 years' life, to watch four hours of TV on a 15" DC powered TV set. In the future, M-Kopa is thinking about a small, highly efficient fridge. In terms of people's demand for electricity services, they want light, TV, phone charging, a fan, and possibly a fridge. All of these services can be competitively supplied with solar, which is more affordable than a grid connection. For such a grid connection in Kenya, a potential customer has to pay a deposit of US$400 to the grid company, and in addition there is a substantial tariff, consisting of a fixed and flexible rate.

After the success of the pilot project, the M-Kopa founders knew that to roll out a proper business they needed additional money. They started a fundraising effort and managed to attract Gray Ghost Ventures out of Atlanta in the United States, who invested $1.5 million in round A, together with a few other institutional investors. This enabled to them to design a back-office that could handle hundreds of thousands of micro transactions every day without a glitch. They also further refined the solar system and its components. Jesse and Chad moved to Kenya in 2011 and started hiring staff. They now employ 850 people and the company has more than 340,000 customers in Kenya, Uganda and Tanzania. After Kenya, they selected Uganda and Tanzania as next markets because the countries all have approximately 40 million people, low electrification rates, and most importantly, have well-introduced mobile money systems. In Kenya, for example, people even buy roadside snacks and pay for them using mobile money. It is expected that in the next 10 years many more countries will introduce mobile money systems. Countries such as Nigeria and Bangladesh have introduced mobile money and its adoption rate is growing fast, so they will be future markets for M-Kopa or other companies using mobile money.

There are an estimated 20 companies like M-Kopa, active in the pay-as-you-go or PAYG segment, which is just about one fifth of all companies in the off-grid solar sector. However, the companies that transitioned from cash-sales to PAYG saw a substantial increase in customer uptake, providing evidence that this is a game-changing addition to the business model.

Solar Businesses

In the last decade, a number of businesses powered by solar energy sprung up all over Africa. With solar becoming more and more affordable, many businesses have added solar panels and batteries to existing diesel systems. This so-called hybridization reduces fuel cost and also reduces the maintenance requirements for the diesel generator, because its use is limited. However, also a number of new business concepts powered 100% by solar energy have been developed.

NICE, the Gambia

In 2001, Dave Jongeneele, founder of Better Future,[14] organized the first European-Gambian management exchange program in The Gambia, one of Africa's smallest countries. Better Future takes leaders and their teams on a journey that aims at opening their horizon and creates an impact beyond their imagination. Paul van Son, then Managing Director Sustainable Energy of Essent, a Dutch utility that is now part of Germany's RWE, participated in that journey. He discovered that many Gambians already had a mobile phone and walked long distances to charge the battery. There were only few, relatively poor functioning Internet shops in the larger communities. Water supply and lighting was also a great problem in many locations. There was some basic power grid infrastructure along the coast, but many communities had no or only very expensive access to the grid. The Gambia has among the highest electricity tariffs in the world. Upon his return back home, Paul founded the Energy4All Foundation to pave the way for decentralized clean energy, clean water and communication/Internet services based on small solar units, wind turbines and batteries. Paul's vision was to build so-called micro-utilities. He quickly found three sponsors: Econcern, Essent, and Rabo Bank. At the time, I was working for Econcern and I got involved. We persuaded Paul to focus on PV and develop an Internet shop concept first as that would lead to quick wins. Econcern founded a company called NICE International (Next-door Internet, Communication and Energy Service shops), with a subsidiary in the Gambia. Energy4All supported that venture and I was elected chairman of the supervisory board. At a later stage, Schneider Electric and the Dutch development bank FMO joined the sponsors.

We developed a concept that would provide Internet and computer services in areas that did not have access to electricity. We started with an extremely efficient design. The selection of all equipment was based on performance and low energy consumption. For instance, our thin client computers used approximately 10 W as compared to 200 W for a typical desktop computer. As a result, a complete NICE-center with 30 computers and a cinema (a large flat screen TV) uses the same amount of energy as one air conditioner! We designed a tracker that had

---

[14]https://www.linkedin.com/in/dave-jongeneelen-b7a2a1.

eight multicrystalline PV modules on it with an overall capacity of 1,840 W. Combined with a battery bank and predominantly DC appliances, the system was able to run completely independent from the grid.

Figure 4.3   The NICE cinema.

The business had three income streams. People that used the Internet paid an hourly fee, just like any other Internet café. On top of that, the centers provided computer classes. And lastly, there was a large flat screen TV that served as a cinema. This was used for educational purposes, but also at night to show soccer matches. The latter was actually important in terms of revenue because it generated additional income from the sales of snacks and drinks.

Figure 4.4   NICE shop, The Gambia.

We started with two centers at two different locations, and although these were operated by NICE the Gambia Ltd, the idea was to roll out the concept as a franchise. The NICE concept attracted considerable attention from other African countries and we looked into introducing the franchise abroad. The management started fundraising for the international expansion in 2008, which looked promising. However, when the financial crisis hit, liquidity dried up very quickly and it was extremely difficult to attract funding for a venture like this. However, we demonstrated it was possible to set up a viable business providing modern ICT services in off-grid areas in Africa, 100% powered by renewable energy. Since then, appliances have become even more efficient, batteries more affordable and solar modules much more powerful and cheaper.

Mini-Grids[15]

The International Renewable Energy Agency (IRENA) describes the difference between off-grid systems and centralized grids in two ways[16]: First, off-grid systems are smaller in size and the term "off-grid" itself is very broad and simply refers to "not using or depending on electricity provided through main grids and generated by main power infrastructure." Second, off-grid systems have a (semi)-autonomous capability to satisfy electricity demand through local power generation, while centralized grids predominantly rely on centralized power stations. As opposed to stand-alone systems for individual appliances/users, mini-grids serve multiple customers. Mini-grids have been in existence for a long time but are traditionally mainly powered by diesel generators. Most clean energy mini-grids are powered by small-scale hydropower, but with the falling cost of solar and battery storage, there is a trend towards hybridizing diesel systems by adding solar.

According to REN21,[17] 2015 saw a lot of activity in the field of clean energy mini-grids in Africa. The company Powerhive[18] secured a loan of $6.8 million from the U.S. Overseas Private

---

[15]Also see Chapter 2.3.
[16]http://www.irena.org/DocumentDownloads/Publications/IRENA_Off-grid_Renewable_Systems_WP_2015.pdf.
[17]http://www.ren21.net/wp-content/uploads/2016/06/GSR_2016_Full_Report_REN21.pdf.
[18]http://www.powerhive.com/.

Investment Corp. (OPIC) to build 100 solar-powered micro-grids in Kenya, which will power about 20,000 households and businesses. Enel Green Power announced that it will invest $12 million for the construction and operation of a 1 MW portfolio of mini-grids in 100 villages. The International Finance Corporation launched a $5 million programmer to develop a market for mini-grids in Tanzania to increase access to energy, while in Mozambique, Energias de Portugal (EDP) secured $1.95 million to finance a 160 kW hybrid solar/biomass mini-grid to power 900 households, 33 productive users and 3 community buildings.

As part of the Sustainable Energy for All initiative (SE4All), initiated by the Secretary-General of the United Nations and the President of the World Bank, which aims to provide universal access to modern energy by 2030, over 50 so-called "high impact opportunities," or HIOs, were identified. One of them is the HIO Clean Energy Mini-Grids, since they can be a viable and cost-effective route to electrification in communities where the distance from the grid is too large and the population density too low to economically justify a grid connection. According to SE4All,[19] Clean Energy Mini-Grids provide an enhanced service level compared with household systems and, depending on local resources and technologies employed, can be comparable to a well-functioning grid.

A scan using the interactive map on the REN21 Web site[20] reveals that there are less than 100 mini-grids with solar PV in operation in Africa, showing the tremendous potential if some of the barriers can be overcome. Those barriers include limited access to finance, lack of entrepreneurs and skilled technicians, and the need for appropriate business models.

Grid-Connected Solar in Sub-Saharan Africa

Structured price-based competitive procurement (auctions) of utility-scale solar PV has been a major factor in driving down the cost of solar PV in the last couple of years. There has been a substantial increase in the number of countries that carried out renewable energy auctions, from 7 countries in 2005 to 60 countries in 2015. The reasons why RE auctions have become the instrument of choice for governments are the following:

---

[19]http://www.se4all.org/sites/default/files/SE4All-HIO-CEMG-Annual-Report-2015-16-2-16.pdf.

[20]http://www.ren21.net/resources/charts-graphs/dre-map/.

- Multiproject auctions typically have lower transaction costs than individually negotiated projects.
- The transparency associated with publicly run auctions will attract international investors.
- Auctions are quicker and less resource-intensive for the buyer.
- Auctions have been proven to yield lower pricing especially in consecutive rounds.
- A bundled approach creates an opportunity to provide stapled finance with better financing terms, reducing the overall cost to the buyer.
- A structured process with clear oversight increases trust, reduces risk premiums, and leads to lower costs.

Countries that have achieved tariffs well below $0.10/kWh through such approach until mid-2016 include Mexico ($0.035/kWh), Jordan ($0.06/kWh), South Africa ($0.073/kWh), Peru ($0.05/kWh), Zambia ($0.06/kWh), UAE (Dubai) ($0.03/kWh), and Jamaica ($0.085/kWh). A number of other countries are now looking into auctions to procure cost-effective solar PV capacity.

Apart from the low cost, another main feature of PV is that it can be built extremely fast. The 75 MW Kalkbult solar PV power plant in South Africa was built in 8 months only, which compares extremely favorably with conventional power. The low environmental impact and modular nature of the technology contribute to this fast deployment.

South Africa

In 2009, the South African government began exploring feed-in tariffs (FITs) for renewable energy, but these were later rejected in favor of competitive tenders.[21] The resulting program, now known as the Renewable Energy Independent Power Producer Procurement Program (REIPPPP),[22] has successfully channeled substantial private sector expertise and investment into grid-connected renewable energy, including solar PV, in South Africa at competitive prices.

The following provides an overview of the first bidding rounds for PV systems:

---

[21]Anton Eberhard et al. (May 2014). *South Africa's Renewable Energy IPP Procurement Program: Success Factors and Lessons*.
[22]http://www.gsb.uct.ac.za/files/ppiafreport.pdf.

   In August 2011, an initial Request for Proposals (RFP) was issued, and a compulsory bidder's conference was held with over 300 organizations attending. By November 2011, 53 bids for 2,128 MW of power generating capacity were received. Ultimately 28 preferred bidders were selected offering 1,416 MW for a total investment of close to $6 billion. Major contractual agreements were signed on November 5, 2012, with most projects reaching full financial close shortly thereafter. The first project came on line in November 2013. A second round of bidding was announced in November 2011. The total amount of power to be acquired was reduced, and other changes were made to tighten the procurement process and increase competition. Seventy-nine bids for 3,233 MW were received in March 2012, and 19 bids were ultimately selected. Prices were more competitive, and bidders also offered better local content terms. Implementation, power purchase and direct agreements were signed for all 19 projects in May 2013. A third round of bidding commenced in May 2013, and again, the total capacity offered was restricted. In August 2013, 93 bids were received totaling 6,023 MW. Seventeen preferred bidders were notified in October 2013 totaling 1,456 MW. Prices fell further in round three. Local content again increased, and financial closure was expected in July 2014. A fourth round of bidding was commenced in August 2014 and concluded with the award of contracts for 1,121 MW renewable energy capacity to six preferred bidders.

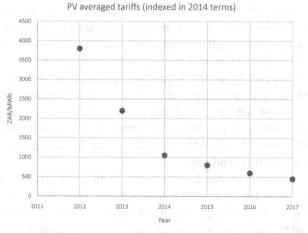

Figure 4.5 Development of electricity tariff in South Africa over consecutive competitive procurement rounds.

Figure 4.5 shows the tariff learning curve achieved over four consecutive rounds of competitive procurement in South Africa.

Similar tariff learning curves have been observed in many other countries in the world, making a strong case for a structured procurement approach with a number of previously announced auctions.

Zambia

Another recent example of a successful solar PV auction is Zambia, where the government worked alongside the IFC.

Of the total installed electricity generation capacity of Zambia of 2,347 MW, hydropower is the most important energy source in the country with 2,259 MW (96%), followed by diesel contributing about 4% to the national energy supply. However, the low rainfall in 2015 has resulted in a national power generation deficit of about 560 MW. Scheduled power outages were having a negative impact on homes and businesses and in 2015 Zambian President Edgar Lungu directed Industrial Development Corporation of Zambia (IDC) to develop at least 600 MW of solar power in the shortest possible time to address the power crisis. The IDC is an investment company wholly owned by the Zambian government, incorporated in early 2014, whose mandate is to catalyze Zambia's industrialization capacity to promote job creation and domestic wealth formation across key economic sectors. The IDC plays its role by serving as co-investor alongside private sector investors. IFC, a member of the World Bank Group, has worked with the IDC on the development of two 50 MW solar PV independent power projects in Zambia, following IFC's recently-launched Scaling Solar initiative. Scaling Solar brings together a suite of World Bank Group services under a single engagement and is a "one stop shop" program. It aims to make privately funded grid-connected solar projects operational within 2 years and at competitive tariffs. It is designed to make it easier for governments to procure solar power quickly and at low cost through competitive tendering and pre-set financing, insurance products, and risk products.

In the case of Zambia, two locations were pre-developed, meaning that the grid connection studies were done, the solar resource was assessed, and a complete finance package was provided by the IFC in the form of stapled finance. They then carried out a two-stage auction, with a pretty strict qualification phase. As a result, a select number of credible developers could focus fully on providing their best price for the solar electricity. The results, which came out in May 2016 exceeded everybody's expectations. The winning bids were for just 6.02 cents per kWh and 7.84 cents per kWh—the lowest prices for solar power to date in Africa, and among the lowest recorded anywhere in the world. Because the tariff is fixed for 25 years and won't rise with inflation, it represents about 4.7 cents per kWh over the life of the project—on par with recent auctions in Peru and Mexico and lower than the lowest tariff achieved for solar in South Africa so far. The winning bidders are expected to complete construction in summer 2017. The two new solar power plants will increase the country's available generating capacity by 5% and will also help to restore water levels in its Kariba and Kafue Gorge dams.

Forecast for Sub-Saharan Africa

According to a recent scenario by the IEA,[23] grid-based electricity generation capacity in sub-Saharan Africa quadruples to 2040. Urban areas experience the largest improvement in the coverage and reliability of centralized electricity supply. Elsewhere, mini-grid and off-grid systems could provide electricity to 70% of those gaining access in rural areas. Considering the abundance of solar energy, the ever-increasing cost competitiveness, and ease of installation, African leaders are now focusing more and more on ambitious renewable energy programs, with a very prominent role for solar PV. With interconnected power pools and future integration of cost-effective storage solutions, a large part of the required new capacity could be solar PV. I am confident that we will see hundreds of gigawatts of solar PV capacity in Africa in the next decades, fueling the continent's economies in a clean and reliable way.

4.2.3 Middle East and North Africa

In the Middle East and North Africa (MENA) region, more than 99% of the population has access to electricity, which is a different situation compared to sub-Saharan countries. Also, several MENA

---

[23]https://www.iea.org/publications/freepublications/publication/AEO_ES_English.pdf.

countries have substantial oil and gas reserves, so the drivers to deploy renewable energy projects are different. Some countries, such as Morocco, import most of their fossil fuels, whereby other countries that have domestic fossil fuels, would like to diversify their energy mix to reduce growing domestic use of own oil and gas. Although for a number of years, announcements about renewable energy projects and policies have been made, so far there has been little real activity on the ground. However, a number of developments point to the fact that finally the turning point has been reached, and the MENA region will start accelerating the introduction of renewable energy. There are areas with a very good wind resource, such as Morocco's Atlantic coast, or Egypt's Zafarana Region, but other areas have only moderate wind speeds. However, the solar irradiation is excellent throughout the region. The recent increasing cost-competitiveness makes for a very compelling case to deploy solar PV at a big scale. In many countries, the demand for electricity keeps growing at rates of between 6% and 8%, and the demand profile, due to air-conditioning, coincides perfectly with the supply of solar energy.

According to the Middle East Solar Industry Association MESIA,[24] more than 2,000 MW of solar power capacity will be tendered in 2016 alone, with an estimated investment of $4 billion. The main countries where that will happen are Morocco, Algeria, the UAE, Jordan, Egypt, Saudi Arabia and Kuwait.

Morocco

Morocco is the only North African country with no natural oil resources and is the largest energy importer in the region, with 96% of its energy needs being sourced externally. The leading supplier of Morocco's energy requirements is Saudi Arabia at 48%. By 2009, Morocco's energy bill had reached $7.3 billion and electricity demand is projected to quadruple by 2030. The government set a goal of reaching 42% of installed capacity (or 6,000 MW) from renewable energy from hydro, wind and solar by 2020, while doubling overall capacity.[25] Solar power in Morocco is enabled by the high rate of solar insolation, about 3,000 hours per year of sunshine but up to 3,600 hours in the desert. Through

[24]http://www.mesia.com/wp-content/uploads/MESIA-Outlook-2016-web.pdf.
[25]http://www.nortonrosefulbright.com/knowledge/publications/66419/renewable-energy-in-morocco.

the Morocco Solar Plan (MSP) it aims to install 2,000 MW of solar capacity by 2020, contributing around 14% of the energy mix in the country's electricity supply. The plan calls for the construction of 5 solar complexes, requiring an estimated investment of $9 billion. This program called NOOR, will be implemented by the construction of solar power plants in Ouarzazate (510 MW CSP and 70 MW PV), Tafilalt and Atlas (300 MW PV), Midelt (300 MW CSP and 300 MW PV), Laayoune and Boujdour (100 MW PV), Tata (300 MW CSP and 300 MW in PV) and solar power plants in the economic zones (150 MW PV). The procurement program is managed by MASEN, the Moroccan Agency for Solar Energy, which is a public-private agency dedicated to implementing the Moroccan Solar Plan and the promotion of solar energy by developing solar power projects, contributing to the development of national expertise and proposing regional and national plans on solar energy.

The Moroccan procurement program has achieved excellent results by awarding power purchase agreements for CSP projects in Ouarzazate, which achieved world record low prices for concentrated solar power at the time. Recently MASEN has launched a dedicated PV program called NOOR PV I and the first phase entails three projects with a combined capacity of 170 MW. The pre-qualification took place in December 2015 and the projects are expected to go online in 2017.

The reason why Morocco's solar ambitions have been more successful than other countries can be attributed to MASEN, which was set up next to the national utility ONEE, and empowered by direct support of King Mohammed VI of Morocco. MASEN has been the main interface with the lenders, such as the World Bank and the European Investment Bank, and has achieved a very favorable financing package, bringing the cost down substantially. In other countries, where the incumbent utility has been a central player in the renewable energy strategy, uptake of renewable energy projects has been considerably slower.

Egypt

Egypt's fantastic history is not only marked by the pharaohs, their pyramids and temples, but one of the earliest industrial solar power plants was built in Egypt. More than 100 years ago, in 1912 in Maedi, parabolic solar collectors were established in a small farming community by Frank Shuman, a Philadelphia inventor, solar visionary and business entrepreneur. The parabolic troughs were used for producing steam, which drove large water pumps, pumping 6,000 gallons of water per minute to vast areas of arid desert land.

In more recent times, Egypt has been struggling to meet its growing power demand with domestic oil and gas. Since Egypt has great solar and wind resources, it was a natural choice for the government to embark on an ambitious renewable energy journey. In 2014, Egypt announced a renewable Feed-in Tariff (FIT) program, with a target of 2.3 GW of solar PV by 2017. Of that, 2,000 MW will be centralized PV power plants, and interest for the program was understandably large. An estimated 1,500 MW of PV projects will start construction in 2016.

The program is supported by multilateral financing institutions such as the IFC, EBRD, and OPIC.

Figure 4.6   The Shuman-Boys solar collector array at Maedi, Egypt, 2016.[26]

---

[26]https://en.wikipedia.org/w/index.php?title=File:The_Electrical_Experimenter,_Volume_3.pdf&page=643.

United Arab Emirates

In 2006, Abu Dhabi, the largest of the UAE's seven emirates, launched the ambitious Masdar initiative. Masdar, which is Arabic for source, aims to build an ecosystem around sustainable energy, enabling Abu Dhabi to remain a central player in the global energy economy. Abu Dhabi has large oil and gas reserves but realizes that over time fossil fuels will be replaced by sustainable energy. Rather than fighting it, or not being part of its development, Abu Dhabi has made the strategic choice to pro-actively invest in sustainable energy technologies, real estate, and projects.

I joined Masdar in 2009 and was responsible, as director of Masdar's clean energy unit, for a portfolio of clean energy investments all over the world, worth more than $8 billion. These included a factory producing PV modules in Germany, factories producing wind turbines in India and Finland, several large concentrated solar power plants in Spain and the UAE, as well as London Array, the largest offshore wind farm in the world.

Masdar's flagship project is Masdar City, a carbon-neutral part of Abu Dhabi close to the international airport. Masdar City is powered by a 10 MW grid-connected ground-based PV system, supported by a 1 MW rooftop system. The 10 MW system is a combination of First Solar and Suntech modules, and the rooftop was supplied by SunEdison. It is ironic that these three companies were all once the largest solar PV companies in the world. However, two of them have had to file for Chapter 11 protection since, which shows that, despite the steady and spectacular growth of solar PV in the last decade, the market development has been rather turbulent and not an easy ride for manufacturers.

After the 10 MW system in Masdar City was built, we started developing Shams I, a concentrated solar power plant with a capacity of 100 MW, to be located in Madinet Zayed, some 140 km from Abu Dhabi. As president of the company, I led the construction, which involved 15,000,000 man-hours. The plant was based on the conventional parabolic trough technology and was supplied by Abengoa. Masdar's co-investors were Abengoa from Spain and Total, the French oil company, which now is the majority owner of Sunpower. The plant was completed in 2013 and has been outperforming its planned specifications ever since. Although the original plan was to build three more CSP plants, the massive price drop of solar PV has changed the immediate priority for Abu Dhabi. The next project will be a 350 MW Solar PV plant, which is planned to go online between 2018 and 2019.

In parallel to Abu Dhabi's efforts, neighboring Dubai, which doesn't have significant oil or gas reserves, started looking at solar as well. The Maktoum Solar Park on the outskirts of Dubai started with a 13 MW solar PV installation in 2014. Initially, Dubai's solar ambitions were relatively modest, but just before COP21 in November 2015, the Dubai Clean Energy Strategy 2050 was announced, which greatly increased the emirate's clean energy targets. Dubai now aims to cover 7% of its power with clean energy by 2020, 25% by 2030 and 75% by 2050. The previous target was 15% by 2030. The main reason for the increased ambition level was the fact that DEWA, the utility company in Dubai, achieved a world record of 5.84 cents per kWh for their 200 MW solar PV project in the second phase of the Maktoum Solar Park. Since Dubai has to import most of its energy, solar energy turned out to be the cheapest form of power in Dubai. DEWA decided to also introduce a rooftop program, based on net-metering. Since, contrary to most other places in the Gulf, including Abu Dhabi, electricity is not subsidized in Dubai, commercial and industrial users of electricity can save money using this scheme. By mid-2016, an estimated 350 MW of rooftop solar PV projects using this scheme are under way, and according to DEWA, the grid can easily absorb 2,500 MW of distributed PV.

But DEWA was in the global headlines again early 2016, when the results of the third phase of the Maktoum Solar Park, a huge project of 800 MW, were announced. The winning bid, offered by my former colleagues of Masdar, came in at a world record low price of 2.99 cents per kWh. The astonishing fact is that this tariff is unsubsidized, mostly using commercial debt. Solar energy is not only the cheapest form of electricity in Dubai, it is so by a wide margin! Of course it helps that the off taker has a solid credit rating, and that loans are still very cheap. However, the same argument is valid for other energy investments.

The original size of the Maktoum Solar Park was increased from 1 GW to 3 GW in early 2015. At the end of 2015 the planned size increased to 5 GW, reflecting the increased appetite for solar PV.

Saudi Arabia

Saudi Arabia has initiated a number of very ambitious renewable energy plans in the recent past. The kingdom produces much of its electricity by burning oil, a practice that most countries abandoned long ago. Most of Saudi Arabia's power plants are highly inefficient, as are its air conditioners, which consumed 70 % of the nation's electricity in 2013. Although the kingdom has just 30 million people, it is the world's sixth-largest consumer of oil. The Saudis burn approximately 25% of the oil they produce—and their domestic consumption has been raising at an alarming 7% a year, nearly three times the rate of population growth. According to a widely read December 2011 report by Chatham House,[27] a British think tank, if this trend continues, domestic consumption could eat into Saudi oil exports by 2021. At this rate of demand growth, national consumption will have doubled in a decade. On a "business as usual" projection, this would jeopardize the country's ability to export to global markets and render the kingdom a net oil *importer* by 2038. Given its dependence on oil export revenues, the inability to expand exports would have a dramatic effect on the economy and the government's ability to spend on domestic welfare and services.

This plain fact has been the main driver behind the government's plans to diversify the energy mix, and the late King Abdullah bin Abdulaziz Al Saud, who passed away in 2015, founded KA CARE, the King Abdullah Centre for Atomic and Renewable Energy by Royal Decree in 2010. KA CARE, based in Riyadh, published scenario in which, by 2032, Saudi Arabia would generate 50% of all electricity from non-fossil fuels. KA CARE's scenario incorporates nuclear, solar, wind, waste-to-energy, and geothermal on the following basis: hydrocarbons—60 GW; nuclear—17.6 GW; solar—41 GW, of which 16 GW will be generated through the use of PV and the balance of 25 GW by concentrated solar power; wind—9 GW; waste-to-energy—3 GW; and geothermal—1 GW. According to their predictions, solar PV will meet total daytime demand year around; concentrated solar power, with storage, will meet the maximum demand difference between photovoltaic and base load technologies; and hydrocarbons will

---

[27]https://www.chathamhouse.org/sites/files/chathamhouse/public/Research/Energy,%20Environment%20and%20Development/1211pr_lahn_stevens.pdf.

meet the remaining demand. This scenario was widely published and created a lot of investors' attention into this promising market, which would be the largest solar PV market in the region. However, 6 years after KA CARE's inception, many things have happened but we are yet to see the promising large-scale initiatives that would really start the market. KA CARE has not yet been fully empowered, which would require a second royal decree, and there seems to be uncertainty which entity will finally lead the efforts going forward. Early 2015 the government said it needed more time to reflect on the best way forward.

However, in April 2016 Prince Mohammed bin Salman, the new Deputy Crown Prince and one of the most influential young leaders in the country, launched the Vision 2030. As part of that vision, the kingdom plans to install 9.5 GW of renewable energy, about a quarter of the previous target. The goals reflect work by the Prince to overhaul the economy of Saudi Arabia, selling off a stake in the state-owned oil company Aramco to diversify away from fossil fuels as a primary revenue source.

With the recently achieved low prices for solar PV, it will just be a matter of time before Saudi Arabia will start deploying solar at a large scale. It makes economic sense and it creates employment opportunities, especially if the kingdom decides to start manufacturing, which is viable because of the market scale. This is an important aspect, since approximately 50% of Saudi's young adults are currently unemployed.

When the market finally kicks off, one of the biggest firms waiting in the wings is Acwa Power International, which is based in Riyadh and owns and operates power and desalination plants in the Middle East, Africa, and Southeast Asia. In the past few years, Acwa Power has signed contracts to produce solar power in several countries—places where the price of conventional electricity is higher than in Saudi Arabia. Acwa won the bid to build a solar farm in Dubai, the 200 MW phase II of the Maktoum Solar Park, which set a world record at 5.84 cents per kWh in 2015.

Although Acwa Power did not win the bid for the next phase, the 800 MW phase III, they came close and are among the most aggressive investors in large-scale solar power projects in the world. As soon as projects materialize in Saudi, experienced developers such as Acwa and others are ready to engage, and transform the fossil fuel giant into a green energy powerhouse.

Finally

The last decade has shown unprecedented growth in Africa and the Middle East, with more people than ever lifted out of extreme poverty. However, the number of people who live in rural areas, where the grid does not reach, has still grown and conventional energy solutions struggle to provide a compelling case for the future. The great solar resource, the increasing cost-competitiveness of solar solutions and innovative business models shine a bright light on Africa's energy future. Large-scale solar farms will feed the grids of African and Middle-Eastern countries, buildings will have solar roofs and energy poor people in the rural areas will have access to affordable and healthy solar technologies that can help them out of poverty. The once neglected solar markets can turn into engines for green economic growth.

4.3 The Present and Future Market in the Americas

Paula Mints[28]

The diverse countries of Canada, the United States, Mexico, and the countries in Central and South America make up the markets for solar in the Americas. Each country or region has different drivers, incentives, import taxes, and degrees of potential for the deployment of solar technologies.

The market in Canada continues to have much unfulfilled potential, but the momentum is spreading even as far as Canada's Yukon Territory. The Yukon, for example, has a population of 38,000 people and recently introduced a feed-in-tariff for homeowners of 21 Canadian Cents/kWh and 30 Canadian Cents/kWh if the PV system replaces diesel.

The United States is a country with high potential for solar deployment and over 50 different markets. The markets in the United States are united by the Investment Tax Credit and divided by different rules and costs for permitting systems, different incentives, different portfolio standards and different rules for net metering with different compensation for valuing solar generated electricity.

Latin America is a region with many different countries all with varying degrees of economic and political stability. Some countries in Latin America have high potential for solar deployment and some, for a variety of reasons, have no potential for solar market development.

Potential is a tricky subject and defining potential is even trickier. Potential for solar deployment is more than complex than DNI (direct normal insulation), or rooftop availability or open land space. Potential for solar deployment is a complex blend of politics, incentives, government support or mandate; affordability, grid reliability, end user tastes and return on investment as well as inflation and the availability of substitutes.

Economic motivation is the basis of the solar decision and overtime the availability of incentives has been the most reliable indicator of solar deployment potential. Incentives have come and gone, often with disastrous consequences for market

---

[28]Paula Mints' biography is on page 292.

participants. For grid-connected solar, the motivation is saving money or making money. The motivations for deploying off-grid solar, particularly concerning rural development, are more complex.

Once the economic motivation is stimulated, the other benefits of choosing to invest in solar such as energy independence and the social benefit of making a choice that is good for the environment support the decision.

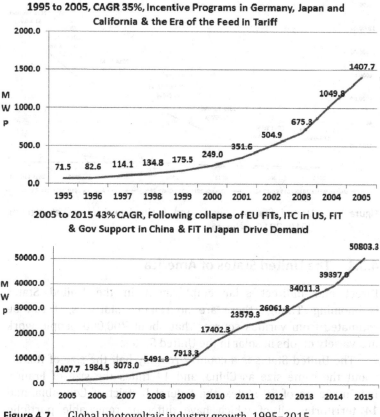

Figure 4.7 Global photovoltaic industry growth, 1995–2015.

Markets can be viewed in isolation, but they are far better considered in context with the global market. The solar industry operates in a global market and all participants interact even when they are unaware of doing so. For example, though the PV system on a homeowner's roof may belong to the homeowner, the kilowatt hours of electricity generated mingle with other kilowatt hours. You may own a grid-connected system, but not the individual kilowatts generated by it.

Figure 4.7 offers global photovoltaic industry growth from 1995 to 2015.

Figure 4.8 presents demand side solar market growth for the United States, Canada, and the countries of Latin America. Specifically, the figure depicts installations in the Americas.

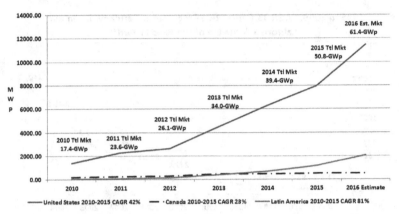

Figure 4.8 United States, Canada and Latin America market growth, 2010: 2016 estimate.

4.3.1 The United States of America

Direct and indirect solar employment in the United States is booming. Though there are no reliable statistics, it can be estimated from various sources that about 200,000 people work in a variety of jobs in solar in the United States.

The United States is a country roughly half the size of Russia, about the same size as China, and 14 times the size of France. It is made up of 50 states, the federal District of Columbia, and 14 territories all of which have different renewable portfolio standards and incentives or lack thereof, and different permitting requirements and costs for the deployment of solar technologies. Currently there is no federal standard or requirement to install solar in the United States. The Obama administration's/EPA's Clean Power Plan (CPP) could be considered a requirement to install renewables, but unfortunately as there are currently over 25 lawsuits against its enactment, it may never have the desired impact. The second pillar of the CPP calls for increased use of natural gas.

Figure 4.9 presents supply and demand for the US PV industry from 1995 through 2015 also indicating the US share of supply (shipments) and demand (deployment) of the total global market place for photovoltaic modules and shipments.

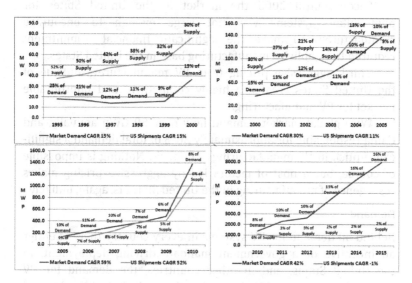

Figure 4.9 US photovoltaic industry growth, 1995–2015.

US Supply

The United States was the photovoltaic manufacturing leader up until the mid-1990s and was a pioneer in the development of crystalline and thin-film photovoltaic technologies. In 1995, the United States supplied over 52% of global modules while consuming 25%. In 1997, the first year that global demand for photovoltaic modules was >100 MWp, the United States supplied 42% of modules while consuming 12%.

Manufacturing of cells and modules in the United States reached its peak in 1995, with its share of global shipments slipping each year thereafter. Currently the United States has 2% of global capacity to produce crystalline and thin film cells. The reasons for slippage are complex.

First, government support for photovoltaic research and development has not been consistent. For example, in 1995 when the United States was the leading supplier of photovoltaic cells and modules, the US congress recommended a 63% cut in funding for renewable energy programs. Another example: in 2007 the Bush administration called for cuts in funding for the National Renewable Energy Lab (NREL). Funding for NREL has been under scrutiny for decades.

Second, until 2005 the market in the United States for photovoltaic systems was not large enough to encourage manufacturers to make the significant financial commitment required to establish a manufacturing base in the United States.

Third, aggressive pricing from PV manufacturers outside of the United States created a highly competitive and unprofitable market for PV manufacturing.

With global market demand now over 50 GWp annually, and China responsible for approximately 30% of total demand a significant uptick in US manufacturing is unlikely. Nothing is impossible of course, but a host of factors led by aggressive pricing makes resurgence of manufacturing in the United States is, again, unlikely.

Demand

Behind all demand for PV systems in every country there is either an incentive or a government mandate driving deployment. In the United States, and in other countries, the demand for PV system deployment is 100% incentive driven, including the demand for residential solar leases. The demand in the United States far outstrips the country's capability to supply it with domestic manufacturing. As previously noted, not even the extension of the ITC,[29] which has stimulated strong growth, is likely to change this paradigm.

The extension of the 30% Investment Tax Credit in 2015 offered stability to a historically unstable market. Renewable portfolio standards, interconnection, and net metering are equally important to ensure that the US market for solar continues growing and, most important, grows in a stable manner.

Solar technologies are installed into applications. Application definitions for photovoltaic installations differ by country or region but in general the application buckets are residential, commercial and off grid. Utility scale is a catchall category the definition of which differs by country. For example, in Japan utility

---

[29]Other financing schemes described in Section 3 (Wall Street and Financing).

scale is >500 kWp. In the United States, residential installations range from 2 to >25 kWp and can be on or off roof. Utility-scale installations are >10 MWp with the average trending larger.

Figure 4.10 offers detail about US application contribution from 2005 to 2015.

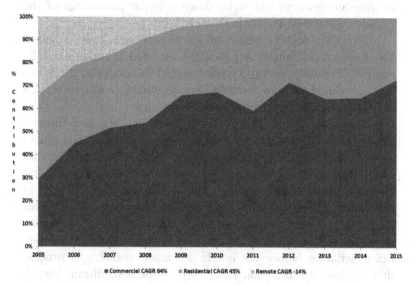

Figure 4.10 US Application Contribution, 2005–2015.

As noted, the United States is not just one market. It is a complex blend of over 50 different markets. California continues as the largest market for solar deployment in the United States though demand in other states is surging.

The residential and business tax credits (ITC)[30] are the most significant financial incentives currently in the United States. Its extension in 2015 added stability to the market for grid-connected solar deployment in the United States. For commercial and utility-scale installations, the ITC remains at 30% for systems that began construction through the end of 2019. The tax credit decreases to 26% in 2020 and to 22% in 2021. In 2023 the tax credit decreases to 10% with no further decreases. Systems that are put into service before the end of 2023 may qualify for their original incentive (30%, 26% and 22%). The Investment Tax Credit (ITC) for residential deployment expires in 2023.

---

[30]Described in Chapter 3.2.2.3.

Modified Accelerated Cost Recovery System (MACRS) U.S. tax code dictates the period of time over which assets can be depreciated. The Modified Accelerated Cost Recovery Systems (MACRS) is a federal standard for the length of asset depreciation. The shorter the depreciation schedule the greater the incentive as investors/owners can write down a larger percentage of the eligible capital expenditure in the near term.

Currently, solar assets can be depreciated over 5 years. The Economic Stimulus Act of 2008 included a bonus whereby eligible renewable energy systems placed in service in 2008 can be depreciated 50% in the first year. The balance is depreciated over the course of the ordinary MACRS schedule.

Clean Renewable Energy Bonds (CREBs) are zero-interest bonds that pay investors returns through tax credits. The power to issue these bonds is vested with the Internal Revenue Service. CREBs can be used by state, local and tribal governments and electric cooperatives to raise capital for the construction of renewable generation and associated transmission assets. This instrument requires significant lead-time to apply for, and then monetize the bonds. Projects must be approved through an open IRS solicitation before the bonds are made available, a process that can take up to a year. This delay can be a significant barrier for projects financed through CREBs.

Loan Programs: The drawback for most energy consumers when considering a PV system purchase is the high upfront cost of installation. This is one reason for the popularity of solar leases. The majority of US states offer loans for renewables.

Interconnection Policies: In order to use the utility grid as a battery, a grid-connected system owner must have an interconnection agreement with the utility. Not all interconnection agreements are created equal and the queue connect can be long, potentially taking years for larger systems.

Net Metering Policies: In 2005 the EPA required public utility commissions (PUCs) and utilities around the country to review net metering and interconnection standards. The development and implementation of standards were left to the discretion of regulators and the utilities (IEEE 1547 was recommended for interconnection). While non-binding, this action did raise the profile of net-metering and interconnection issues, which is critical for grid-connected systems. Predecessor to net metering, the

Public Utility Regulatory Policies Act (PURPA) of 1978[31] mandated that utilities were to procure electricity from qualified non-utility power producers at the utilities avoided cost rate. This avoided cost, typically the fuel costs incurred by a traditional fossil fuel plant, is insufficient to make solar power projects viable.

The ability to net meter is as crucial as interconnection to the continued deployment of distributed generation (DG) residential and commercial PV systems in the United States. Most utility net metering programs have limits. These limits, of course, constrain the potential of grid-connected PV system deployment.

With net metering the utility either credits or pays for the electricity that is fed into its grid. In some cases the utility absorbs the excess electricity, in some cases the utility rolls the excess over into another period and in some cases the system owner is paid a set rate for the excess electricity (similar in theory to a feed-in tariff). Different utilities have limits as to how much electricity they will allow to be net metered.

Utilities across the United States are pushing back on net metering by ending programs (Hawaii), adjusting how reimbursement for net excess is calculated, and adding additional fees. Many utilities are pushing to have changes made retroactive. A switch in billing structure in California to Time of Use and a new peak changes the economics for homeowners with PV systems on their roofs, even for legacy system owners.

From 2005 to 2015, the residential application in the United States grew at a compound annual rate of 53%. Though net metering is only one driver of this growth, it certainly makes the economic case for the homeowners particularly when net excess electricity is credited at retail rates.

Utilities did not expect solar industry growth to accelerate so significantly and there is no doubt that they see this growth in terms of revenue decay.

Currently, and it must be stressed that there is no clear trend in terms of outcomes, the following changes to net metering are being sought state by state and case by case:

- Additional or increased fees for net metered systems: Depending on the fee this can dissuade potential buyers/lessees and high fees can upend the economic benefit for buyers/lessees

[31]Also discussed in Chapter 2.1.

- A switch to Time of Use rates: Higher prices for electricity during peak times and lower payment for net excess during off peak times can upend the economic benefit for buyers/lessees
- Lowering the reimbursement for net excess to avoided cost: Danger of undervaluing net excess and upending the economic benefit for buyers/lessees
- Changing the rules for reimbursement for net excess: A blast from the past that could (in the worst case) result in the net excess being granted to the utility
- Making all of the above retroactive

The utility argument for altering how net excess is compensated and for adding additional fees is economic. Utilities argue that ratepayers with solar systems (leased or owned) are renting less electricity from the utility and thus not paying their fair share for overall maintenance. The argument continues that the costs are unfairly shifted to ratepayers without solar systems on their roofs. Establishing a fair fee for solar customers over and above the base fee all ratepayers pay is not simple. The addition of fees for solar customers should not be overly punitive or appear as a referendum against DG solar. After all, ratepayers without solar systems benefit from the clean energy generated by ratepayers with solar systems. Also, the electricity future likely includes more self-consumption and more micro-grids as well as a new operating and revenue model for utilities.

Property Assessed Clean Energy (PACE) Financing Policies: PACE is an assessment on a residential or commercial property, essentially a lien but similar to a loan, that provides a vehicle for a homeowner (traditionally) to install a PV system and repay the lien over a period of years as part of their property tax bills.

Rebates: Simply, a rebate pays a system owner a certain amount per kilowatt installed. In the United States, there are 50 states, 1 federal district, and a number of independent territories or tribal reservations. While federal powers supersede those of the states and other territories, there are still state and local tax codes, political assemblies, local government budgets, and regulatory frameworks that interact with, or modify, federal authority. States also exercise significant regulatory control over utilities within their jurisdiction, typically through the state public utility commission, (PUC).

Renewable Portfolio Standards: An RPS is a mandate that requires the state's publically owned utilities to install a percentage of renewables in their territory, or, buy renewable generated electricity. RPS standards vary state by state. Rebates and other incentive programs are often set up to aid the utility in meeting its goal. The punitive actions a utility faces for noncompliance vary from state to state. A few states have recently eliminated or reduced their RPS requirement, but many states are expanding their RPS, in some cases significantly.

SRECs[32]: Solar renewable energy credits are tradeable instruments (similar to stock certificates). One SREC is equal to 1,000 kWh. The value of an SREC is set by the market, that is, if there is an oversupply the value falls, if there is an undersupply (shortage), the value rises. To be useful a market should have an RPS with a punitive measure severe enough to encourage compliance and the system owner must be able to own and trade the SREC or, solar renewable energy credits. The future of Renewable Energy Credits as a system-financing tool cannot be ignored, though as the market sets the REC price, its effectiveness to stimulate the residential and commercial market is questionable. The problem with the RECs is that the behavior is similar to that of the stock market, that is, an oversupply of RECs renders the value at a low level insufficient to stimulate the market. Renewable Energy Credits can function on their own or as a compliance tool. There is no standard protocol for defining the environmental attributes of a REC. There is no consensus as to the volume of avoided greenhouse gasses or other pollutants. Currently, this is not a significant concern, but how these attributes are interpreted could affect other markets, such as sulfur emission allowances or carbon credits.

MATOC (Multiple Award Task Order Contract): In 2012 Congress mandated that the US Military through the DOD spend $7 billion for R&D as well as deployment of 3 GWp of RE. The requirement is 22% by 2020 and covers all RE technologies.

Cap and Trade: Cap and trade plans can be considered a renewable technology driver as these government-mandated programs encourage (though do not demand) switching. The plan to reduce emissions by attaching an economic incentive for doing so by setting a cap on allowable levels of emissions and

---

[32]SRECs are also discussed in Chapter 3.2.2.2.

allowing industries to trade for carbon allowances. In the United States, there are state as well as federal (EPA) cap and trade programs.

Business Models

Residential and PV system sales are typically financed and rarely, though this does happen, cash purchased. For residential purchases home equity financing is common. PACE is a method of financing that allows loan repayment through the homeowner's property taxes. The interest rate for PACE financing has not proven attractive to homeowners. Commercial PACE, somewhat new in the United States, is slowly gaining traction.

A power purchase agreement or PPA is a long-term agreement to buy electricity from a specific company's solar system[33] PPAs, basically, involve a seller, typically an independent power producer (IPP) but behind which are investors, EPC, installers, system integrators and others, a host (land or roof space is either leased or acquired), and a buyer (typically a utility). The larger the system (the more megawatts involved) the more complicated the structure and the more investment necessary all towards mitigating risk.

Participants in the PPA model in the United States, and with the decline of the FiT model, globally, are extremely sensitive to component price changes in both directions, and must manipulate multiple incentives to achieve healthy margins, while working under sometimes-dire financial conditions.

The residential lease/PPA has expanded in recent years to include deployment into small to the mid-market commercial sector. Under the lease model, the lease operator funds the installation and collects a monthly lease payment. The lease payment includes an annual escalation charge. Residential solar lease or residential PPA providers charge a monthly fee along with assuming the rights to the incentives, including tax incentives such as the ITC, and RECs (renewable energy credits, which are tradable instruments). Older installations fund ongoing construction under this model. A leased PV system on a residential roof is not typically an asset in the sale of a house.

---

[33]See also Chapter 3.3.4.1.

Customers of the residential PPA are responsible for all the electricity generated from the system on their roof and not simply the electricity that they use. At the end of a period of time, typically annually, the utility calculates the difference between the electricity the system produced and the electricity the homeowner uses and either sends a bill or a refund. This process is referred to as the "true up."

Energy Yieldcos[34] are publicly traded companies formed to own operating energy assets and use the proceeds to fund future deployment. Yieldcos are a recent, growing and primarily US phenomenon for ownership of renewable energy assets, though this instrument is not new to the conventional energy industry. This concept assumes that the returns from the project portfolio will be stable enough to support future deployment. In philosophy this is similar to the residential solar lease, which relies on assets in operation to fund future deployment.

Community solar, or group buy, is not a new business model and is a grid-connected application similar in theory to the village grid off-grid application. This business model falls into the commercial application. Under this model, individuals or companies that either do not wish to own a system or have an inadequate roof or space to install a photovoltaic system can buy into a ground or rooftop installation and own a panel or panels. Participants, typically residential, receive a credit on their electricity bill and can transfer ownership more easily than can residential solar lease or power buyers. A community solar installation can be deployed on parking structures and rooftops but are more commonly deployed in ground mount configurations. This model could prove risky to investors if the community solar operator fails and abandons the system.

Today, Tomorrow and the Future of US Solar

Despite the unprofitability of the residential solar lease, attacks on net metering by utilities, and underbidding on PPAs, PV deployment in the United States is currently trending on an accelerated track and the extension of the ITC will continue. Figure 4.11 offers a forecast from three scenarios: low, conservative, and accelerated.

---

[34]See also Chapter 3.5.2.

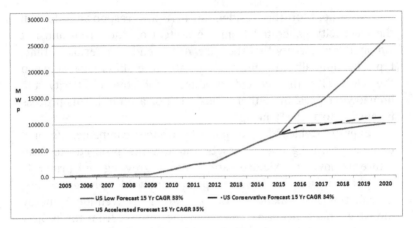

Figure 4.11 US demand forecast, 2015–2020.

The outlook for PV manufacturing in the United States is not as rosy. At the beginning of 2016, the United States had cell capacity of 1.4 GWp and module assembly capacity of 2.7 GWp. Exuberant announcements for capacity additions have been delayed. Despite its strong and potentially stable market, it is unlikely that the United States can increase supply share from its current 2% of global supply.

4.3.2 Canada

The United States has close to 200,000 people employed in its sector and Canada has about 35,000 people employed in its solar sector. At the beginning of 2016, Canada has no crystalline cell or thin film manufacturing capacity and approximately 1.1 GWp of module assembly capability. When Canada announced its feed-in tariff in 2009 there were expectations of a multigigawatt market for PV installations. For a variety of reasons, including inadequate transmission, significant decreases in FiT rates as well as changes to the programs structure, Ontario has never lived up to its initial hype.

Canada is a large country consisting of 10 provinces and 3 territories. Oil and gas are the country's economic drivers and Canada is the largest provider of natural gas to the United States, though the fracking boom in the United States threatens the country's source of revenue in this regard.

All provinces in Canada have net metering. Alberta has a Carbon Tax program. Alberta also has five trillion cubic feet of natural gas reserves and is the world's second largest exporter of and fourth largest producer of natural gas. Alberta also has 39% of Canada's total oil reserves.

The market for PV installations in Canada, specifically Ontario, has been steady but not explosive. The original design of Ontario's FiT did not take into account the poor transmission infrastructure and megawatts of projects were delayed and finally canceled. Yearly alterations to the FiT have led to confusion and dissatisfaction. A domestic content requirement was doomed from the start due to the uninspiring level of activity, that is, a market needs to be multigigawatt level to encourage manufacturing participation.

Following the WTO ruling on the illegality of its domestic content requirement, the government of Ontario has removed it, effective mid-2014. In reality, given the dearth of PV manufacturing capacity in Canada and the relatively slow progress of its FiT, the domestic content requirement was ineffective in promoting manufacturing in Canada.

4.3.3 Countries in Latin America

The market for flat-plate PV deployment in Latin America is accelerating driven by electricity restructuring in Mexico as well as high demand in Chile, and the Caribbean. In many other countries, the demand for solar deployment is constrained by high import taxes, low incentives, high levels of crime and violence and economic vulnerability.

Latin America does not offer the robust incentives available in other regions though individual countries offer tax incentives and other encouragement. Value added taxes and additional tariffs imposed on imports can increase the price of a module significantly in many countries. In general, plans to install solar are pursued via tenders and PPAs, though Argentina has a Feed-In-Tariff.

Renewable electricity auctions throughout Latin American countries have delivered good and bad news. In good news, solar (PV) has won in terms of gigawatts. Unfortunately, underbidding undervalues PV-generated electricity and sets a precedent in terms of valuing solar generated electricity. In Mexico, its first auction offered 1.7 GWp for wind and solar. Sixty-five percent, or

1.1 GWp, was awarded to PV projects with bids from <$0.04/kWh to ~$0.06 kWh. In Chile the value of solar generated electricity on the spot market is sometimes zero.

Persistent roadblocks to deployment in Latin American countries include transmission, delays due to feasibility studies and paperwork, project cancellation, economic upheaval, high import taxes, low tender bids and slow interconnection approval and a preference in many countries for mini-hydroelectricity.

In many countries, urban inhabitants still pay less than the cost for grid electricity but the electricity reform in many countries is leading to higher electricity rates. High import tariffs in many countries increase the cost of importing solar technology and therefore the cost of installing it. Due to its low labor costs and the assumed high potential of its market, not to mention (in the case of Mexico) proximity to the United States, Latin America has become an interesting region to set up module assembly, though many countries import modules from the United States and Canada. High expectations for market growth and continued deployment are often disappointed.

Delivery systems for rural electrification programs are through local dealers/distributors, local franchises, and rural cooperatives. In some cases, PV manufacturers have remained committed to rural electrification in these countries. However, for these developing areas, installations must be subsidized, and users must be educated and their expectations properly set. Reasonable tariff structures must be put into place in order for these populations to accelerate up the learning curve, thus reducing the cost of future projects in the area.

There remains a strong need for electricity among this region's rural poor, along with off-grid agriculture and solar home systems for the rural middle class. General barriers are the expense of the technology, ability or inability to pay, unavailability of financing, poor after-sales service (maintenance) and little or no training into system maintenance, poorly set expectations, theft of component parts, compounded by inadequate government programs. There continues to be interest in providing grid-connected systems to rural populations, unfortunately, this also remains unlikely in the near term.

Difficulties implementing rural electrification programs in Latin America are similar to those experienced in other areas of the developing world. In some cases, end users inherit the problems and mistakes of the project coordinator and those implementing the project. These problems, essentially external to the user in that they are out of the user's control, include incorrect site selection (in the case of water pumping and other village power applications), equipment failure (usually the water pump, battery, etc., and not the module), and inadequate installation (affecting the entire system). Often the intended recipients of rural electrification projects do not understand what to expect from the technology and are thus disappointed (poor expectation setting). The beneficiaries of the project may also see a rural electrification program as second-class in comparison to what the urban grid-connected population receives. However, the high upfront cost of the technology remains the principal barrier for the extremely poor populations for whom solar electricity may be the only answer for an improved quality of life. For these populations, need must be balanced with some financial mechanism, including tariffs, rental fees, a down payment, user fee, or credit to finance the necessary ongoing training and maintenance.

4.4 The Present and Future Markets in Europe

Paula Mints

Europe led the great feed-in-tariff revolution from which the global solar industry has an ongoing hangover and from which it may never fully recover. In 2008, 2009, and 2010, Europe was 80% of global demand making the region highly attractive for manufacturers, investors, and project developers. During the height of the feed-in-tariff period, expectations for extreme price and cost decreases were set and pegged, unfortunately, to aggressive and artificial prices for modules imported from China. During the heyday of its strong market, participants believed that the region would remain the largest and most influential global market. Europe is no longer the largest market but it does remain influential offering lessons about market behavior.

Lesson 1: Demand for grid-connected solar deployment is driven by government mandate and/or incentives.

Lesson 2: Incentives drive a particular type of live-for-today market behavior in that participants cannot be expected to line up in orderly fashion to accept an even portion of available demand.

Lesson 3: A region's manufacturers will not necessarily thrive or profit from an available incentive.

Lessons 4: No matter how carefully the rules are constructed, people will look for ways to get around the rules and they will find them.

Finally, markets rise and markets fall in all industries.

Figure 4.12 presents select global markets in transition, capturing the period when demand into Europe dominated.

Throughout the height of its FiT-driven demand, manufacturers in Europe were, in the main, unable to take advantage of its strong market. PV cell and module manufacturers in Europe had the leading share for shipments in 2007 and 2008. Figure 4.13 presents demand (market) and supply (PV module) growth for Europe from 2005 to 2015.

Europe's overall market was strong during the FiT era, remaining strong even through the great global recession of the late 2000s. Market behavior from 2005 to 2015, which included the rise and collapse of several markets, is linked to economic factors, but primarily to the availability of FiTs as well as the generosity of, degression of and the imposition of retroactive changes to FiT programs.

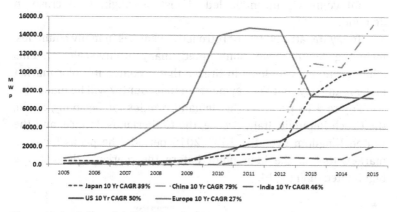

Figure 4.12   Select global markets, 2005–2015.

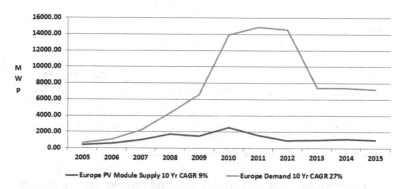

Figure 4.13   Europe PV manufacturing and demand, 2005–2015.

In the period before the great global recession, banks and other investors did not require performance guarantees and poorly designed systems and poorly assembled module product crept into Europe's market. Countries in Europe with FiTs underwent abrupt and in some cases retroactive changes to the rules and the tariff rates. These abrupt changes shook investor confidence and drove down IRRs, specifically, with retroactive changes returns that were assumed to be stable abruptly became unstable. For example, a retroactive tax established in the Czech Republic led to a market crash with no expectations for recovery, while changes to the amount of electricity that would be reimbursed in Spain (as well as other countries) along with the abrupt cessation of that country's incentive led almost overnight to a crash in demand.

Today the aftereffects of retroactive changes in many countries, including Spain, have bankrupted many utility-scale systems while economic problems in Greece that predated its solar market make it unlikely that FiT payments in that country will be made.

Figure 4.14 presents the demand profiles for Germany, Spain, the Czech Republic, Italy, Greece, the Netherlands, France, and the United Kingdom over the 2005–2015 period. The peaks for each market, that is the high point before the market's decline, are highlighted in Fig. 4.14.

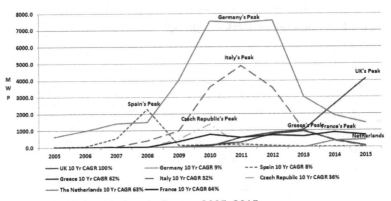

Figure 4.14 Select markets in Europe, 2005–2015.

As previously noted, Europe's cell and module manufacturers did not have much of an opportunity to enjoy the region's strong market. Inexpensive module imports from China created a highly competitive market, or more appropriately, a highly uncompetitive market.

During the mid to late 2000s, PV was a job generator for many countries in Europe creating installation, development and manufacturing jobs. Unfortunately, inexpensive imports flooded the markets in Europe and as incentives decreased, ended and changed retroactively the market slowed in many markets, crashed in others, and despite surges in demand in select country markets, many of the jobs created during this time have been lost.

Figure 4.15 presents PV market growth globally and in Europe from 2005 to 2015. In 2007, photovoltaic deployment in Europe was 71% of global demand. In 2008 Europe had an 83% share of global demand and in 2010 the market in Europe was 80% of global demand.

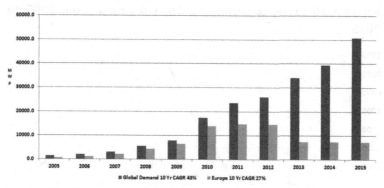

Figure 4.15 Europe and global PV industry growth, 2005–2015.

The markets for PV systems in Europe offers lessons about how to successfully stimulate demand and what happens when incentive schemes go disastrously awry. Germany stands out as a success story, while Spain with its overgenerous program and retroactive changes highlights the need for care when designing market-stimulating incentives.

Germany

Germany was a pioneer in the promotion and support of photovoltaic technologies beginning in the late 1990s. The country's historic Renewable Energy Law, (EEG) was instrumental in altering the incentive paradigm globally. This influential incentive model is responsible for stimulating significant demand. Germany proved that by providing the appropriate incentive, a market for photovoltaic (and other solar) installations could be established and grow.

In 1998 Germany launched its 100,000-roof program administered by the KfW Development Bank. This program granted more than 60,000 loan applications by the time it ended on June 30, 2003. Following the introduction of higher feed-in tariffs in 2000, the number of loan applications increased sharply. Approximately 20,000 loans were granted in 2003.

Germany's feed-in-tariff incentive scheme served as the model for feed-in-tariffs in other European countries, serving as a base for the country's strong demand until 2012. Currently Germany's government favors self-consumption as a driver with incentives for storage and tender bidding for ground mount installations. Figure 4.16 presents Germany's demand profile from 2005 to 2015.

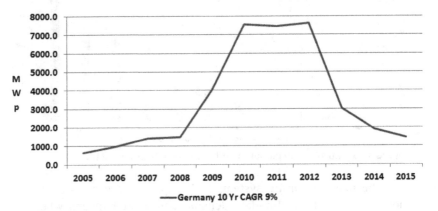

Figure 4.16 Germany demand profile, 2005–2015.

The United Kingdom

Historically and until recently, the market in the United Kingdom was quite small. In 2003, PV installations in the United Kingdom increased by 88% from 1.6 to 3 MWp, primarily as a result of research and grant programs under the UK Department of Trade and Industry's (DTI) New and Renewable Energy Program. This three-year program began in 2003, funded at €20 million (about US$24 million) and administered by the Energy Saving Trust finally came to an end in March 2006. The program funded the residential and commercial applications, and also retrofit and BIPV installations. The program was not considered a complete success as it did not result in the expected system cost reductions.

Almost immediately following its announcement, the UK feed-in tariff began to experience structural changes that were implemented in an attempt to avoid the overheated markets observed in other countries. Overall, despite starts, stops, changes, and internal battles, the United Kingdom's FiT and Renewable Obligation programs had a rocky start, but it was finally successful at jumpstarting what had been a stagnant market for solar.

Some history: In 2005, the Energy Saving Trust began to think in terms of medium- and long-term solar potential, out to 2050. This rethinking of priorities was expanded in 2007 with the Climate Change and Sustainable Energy Act, which tackled specifics such as retrofitting also broaching the potential of PV as a significant electricity generator. There was more positive momentum in 2008 with the Code for Sustainable homes, which included a goal of carbon neutrality in new building by 2016, followed by the renewable energy white paper in 2009, and Energy Act, that called for 5% energy generation by RE technologies (electricity and heat) by 2020.

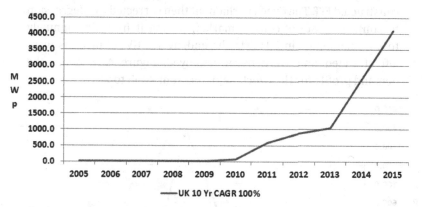

Figure 4.17 UK demand profile, 2005–2015.

When the UK FiT was implemented on April 1, 2010, there was every reason to expect a successful and sustainable market and in fact, 68 MWp was installed within months. Less than a year after the successful launch of the United Kingdom's first FiT, the government underwent an emergency review and in order to discourage the domination of the tariff by solar farms (utility scale), it enacted significant cuts for systems >50 kWp. This rapid turnaround introduced instability into the tariff market and investor insecurity and left many fledgling solar businesses with worthless sales pipelines. In the United Kingdom, the FiT was defined as a tax forgone, so, even if it was a levee, it was still a tax. The FiT was also included in an overall government spending limitation and the amount to be spent on it was been capped.

Since May 2015, cuts and other changes to solar programs, including the small-scale FiT, have been made and the renewable obligation for projects ≤5 MWp closed as of April 1, 2016. It seems as if the UK market for PV installations has hit its peak. Figure 4.17 presents the demand profile for the UK market from 2005 to 2015.

The Czech Republic

The market for solar electric systems in the Czech Republic exploded in 2009, rising from single digit megawatts to 536.8 MWp. The country's market then ended abruptly in 2010 at ~1.4 GWp of demand, entirely because of a too generous and poorly constructed FiT. The Czech parliament then corrected the degression to control its out-of-control market, raising it from 5% annually to 25% beginning in 2011. In the end, over 6 GW of projects were left stranded and were never deployed. Figure 4.18 offers the demand profile for the Czech Republic from 2005 to 2015.

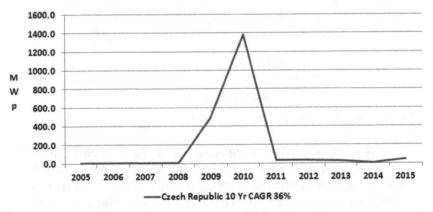

Figure 4.18 The Czech Republic demand profile, 2005–2015.

Spain

Spain's experience with its solar FiT should serve as an object lesson to the global solar industry. When Spain introduced its extremely generous FiT in the mid-2000s the country had an unstable economy, a housing crisis, and high levels of debt and unemployment. Almost immediately after its introduction, Spain's generous FiT stimulated accelerated deployment of CSP, CPV, and flat-plate PV installations.

In 2008, Spain was one of the fastest growing strongest markets in the world. Unfortunately, the market expanded too fast leaving poor-quality systems and rapacious speculation in its wake. Retroactive changes to FiT rates as well as the amount of electricity that could be compensated led to bankrupt installations. Royal decree 413/2014 retroactively capped earnings from systems installed under the country's FiT. Currently, there are several developer-led lawsuits. Figure 4.19 offers the demand profile for Spain from 2005 to 2015.

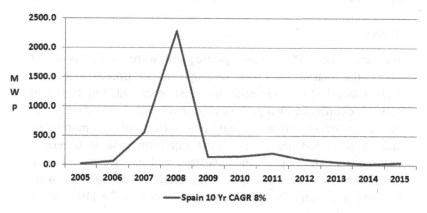

Figure 4.19 Spain demand profile, 2005–2015.

Italy

Italy provides another object lesson for the global solar industry, that of a problematic market instituting an insupportably generous FiT and attempting to control it after the fact. Despite administrative roadblocks and high government debt levels along with a significant need to upgrade its transmission system, Italy was once one of Europe's (the world's for that matter) strongest markets for PV. Other problems included long queues for permits (in some cases, systems waited ~2 years for permits, a situation that effects cash flow), and transmission problems in southern Italy. In 2014 Italy's government instituted retroactive limits on FiT remuneration. Figure 4.20 offers the demand profile for Italy.

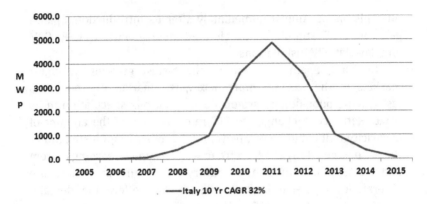

Figure 4.20   Italy demand profile, 2005–2015.

Greece

For many years, PV industry participants were highly optimistic about the market for solar deployment in Greece and from 2011 through 2013 this optimism was rewarded. Unfortunately, economic concerns, changes in government and austerity measures forced on Greece brought a rapid end to the country's market for solar deployment. By far the most significant risk in Greece is the risk that developers and investors will not see a return on investment because the FiT remuneration will not be paid. The country's unstable FiT has been limited to 20 MWp of PV deployment annually.

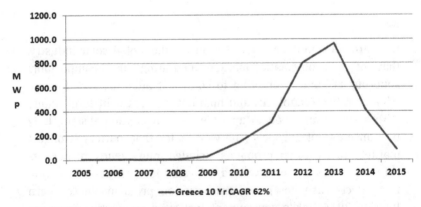

Figure 4.21   Greece demand profile, 2005–2015.

France

France is one of the few countries in Europe, or the world, to attempt to reward the deployment of building integrated PV. Unfortunately, as with other countries in Europe, France struggled to control its FiT-driven market. In 2010 the government of France, concerned over apparent growing speculation in its market, reduced the feed-in tariffs for electricity from roof-mounted systems by 7% and made the reductions retroactive to November 1 of the previous year for systems that did not meet specific criteria. At the time the government stated its concern over a speculative bubble that developed in November and December 2009 in advance of an expected FiT decline. A significant number of applications were filed in November 2009 and December 2009 and the government grew concerned about the effect on the country's utility ratepayers. Other concerns included poor-quality installations that required systems, in some cases, to be un-installed. In the late 2000s, France also had a shortage of qualified installers.

Some FiTs created bubbles because of too high tariffs and few rules, and some created bubbles because of too high tariffs and unwieldy structures. France's FiT fell into the latter category. France's initial goal with its FiT was to encourage a strong market for BIPV. Instead, the market for commercial ground mounted systems and retrofit rooftop systems grew rapidly, with true BIPV growing much more slowly. In 2011, France changed the definition of BIPV so that only systems that were part of the building weather bearing façade were considered BIPV. Other building installations were considered simplified BIPV and retrofit systems were not considered BIPV.

France's dependence on nuclear energy is the single most limiting factor to accelerated growth of PV in the country. On the order of 75% of the country's electricity is generated by nuclear, electricity rates are relatively low and France is the world's largest net exporter of electricity. France's government is determined to reduce its dependence on nuclear and is set to reduce nuclear energy's share of electricity production by 50% by 2025. Decrease in the use of nuclear has not made France an entirely solar-friendly country. France's first FiT favored BIPV under a convoluted and difficult-to-parse scheme that did not stimulate a market for BIPV deployment. France has increased its solar deployment goals from 500 MWp a year to 1 GWp annually.

Figure 4.22 offers the PV demand profile for France from 2005 to 2015.

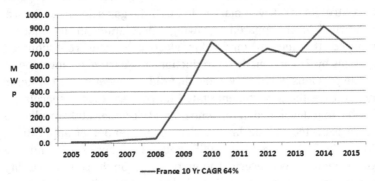

Figure 4.22 France demand profile, 2005–2015.

The Netherlands

Solar deployment in the Netherlands was modest until 2013. In 2013 the Agreement for Sustainable Growth was established requiring modest goals of 14% electricity generation from renewable technologies by 2020 and 16% by 2023. Modest goals aside, the government implemented a feed-in premium for large-scale deployment. The premium, the SDE+ is the difference between the market cost of conventional energy generated electricity and electricity generated by PV. When the cost of conventional energy is high the SDE+ is low and when the cost of conventional energy is low, the SDE+ is high.

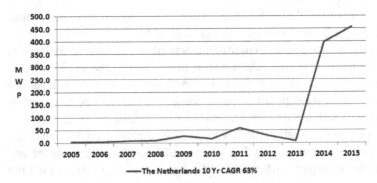

Figure 4.23 The Netherlands demand profile, 2005–2015.

Residential and small commercial deployment in the Netherlands can avail itself of net metering. Loans and tax benefits are available for PV deployment, including by collectives, that is, neighborhoods and communities. Figure 4.23 depicts PV deployment in the Netherlands from 2005 to 2015.

The Future of PV Deployment in Europe

Observing the market behavior in Europe offers a valuable lesson for all flat-plate, CPV, or CSP solar industry participants. Boom and bust cycles abound. Erratic market behavior, as indicated in Figure 4.14, was primarily driven by poorly constructed incentive schemes and a lack of understanding about market behavior as it applies to incentive driven industries.

Despite it all, the European feed-in-tariff incentive, pioneered by Germany, was the most successful economic instrument at stimulating the demand for solar deployment. The European FiTs drove the global demand for solar deployment to multigigawatt levels, awakened worldwide appetite for clean, solar generated electricity, and encouraged innovation in crystalline and thin film R&D, module assembly and installation techniques. Multimegawatt installations, referred to as utility scale, became an established trend because of the feed-in-tariff incentive.

Even a global recession could not completely disrupt the solar-momentum in Europe.

That the market in Europe is softening is a fact. Tender bidding and net metering schemes will not replace generous incentives. The market in Europe is settling into lower but stable demand that should ensure a steady market going forward.

4.5 The Present and Future Markets in Asia

Paula Mints

Overview

The global solar industry always seems to be going through growing pains. Markets emerge, mature, and slow, sometimes ending abruptly. Leading manufacturers come and go sometimes failing entirely. Prices rise and prices fall. Since the late 2000s, both manufacturing and demand leadership have been shifting to the countries in Asia.

The countries of Asia, Southeast Asia, and West Asia currently have 95% of the global capacity to manufacturer crystalline, and thin film cells and several countries in these regions consume 60% of module product. China has been the leader in annual shipments of photovoltaic modules since 2009 and with its new 1 GW goals for CSP appears primed to establish leadership in the concentrating solar sector.

The countries in Asia include Japan, China, South Korea, and Taiwan. The countries in West Asia include India, Nepal, and Pakistan. The countries in South East Asia include Indonesia, Vietnam, Malaysia, Thailand, and the Philippines.

Figure 4.24 presents the supply and demand shares for the countries in Asia, West Asia, and South East Asia for 2015 as well as the share of all other countries. Supply refers to shipments or, specifically, to the manufacture and sales of photovoltaic (PV) modules.

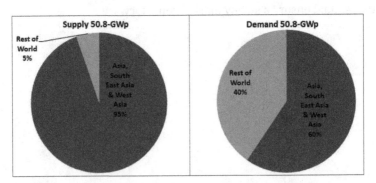

Figure 4.24 Asia, South East Asia, and West Asia share of 2015 supply and demand.

From 2005 to 2015, the markets for PV installations in Asia, Southeast Asia, and West Asia combined grew at a compound annual rate of 50%, faster than the market as a whole. Figure 4.25 presents PV market growth globally and in Asia from 2005 through 2015.

The markets for several countries in the Asia(s) began emerging in 2010. Figure 4.26 shows the demand profile for the rapidly emerging markets of India, Thailand, and Pakistan as well as the dominant markets of China and Japan from 2005 to 2015.

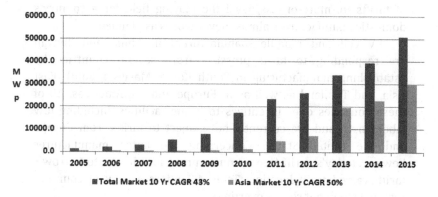

Figure 4.25   Global and Asia market growth, 2005–2015.[35]

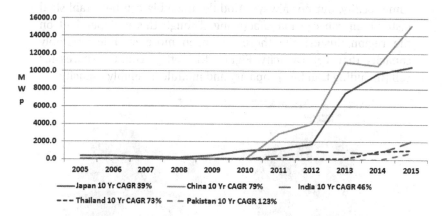

Figure 4.26   Select markets in Asia, 2005–2015.

---

[35]CAGR, compound annual growth rate.

Supply

Overtime, supply leadership in the PV industry has changed, switching from country to country for various reasons. Price leadership, referring to the lowest price, is the dominant theme. Price domination fueled a shift to China as the supply leader in 2009.

Currently, another shift in manufacturing is taking place, driven by tariffs, that is, the type of tariff that penalizes an activity instead of feed-in tariffs that reward activities. The use of tariffs to, more-or-less, level the playing field for a country's domestic manufacturers almost never works as planned.

PV cell and module manufacturers in China and Taiwan are responding to tariff pressure from several countries by establishing manufacturing in South Korea, Malaysia, South East Asia, and Thailand as well as in Europe and the Americas. All of these countries offer incentives to locate facilities, including low labor costs, and are currently not subject to tariffs from Europe and the United States. Concerning the latter, manufacturer beware, today's tariff-free manufacturing location is tomorrow's tariff-scarred tariff location. Table 4.2 offers a view of country/regional shipment shares overtime.

Cell capacity and module assembly capacity are often in the same facility, but not always. Module assembly can be established faster than can cell manufacturing. Though this is dependent on the region, historically there has been more module assembly capacity than cell capacity. Figure 4.27 offers country shares for cell (including thin film) capacity and module assembly capacity.

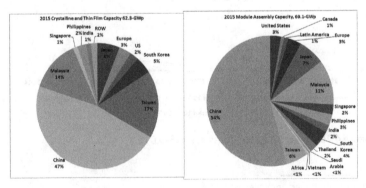

Figure 4.27 Cell and module assembly capacity, 2015.

Table 4.2 Country cell/module shipment shares 2005–2015

Year	US. Share	Europe Share	Japan Share	Philippines Share	Singapore Share	India Share	South Korea Share	ROW Share	China Share	Taiwan Share	Malaysia Share	Ship. MW	Average selling price
2005	9.5%	28.9%	50.7%	1.4%	0.0%	0.9%	0.0%	1.1%	4.2%	3.3%	0.0%	1,407.7	$3.03
2006	6.9%	30.8%	44.5%	2.3%	0.0%	0.0%	0.0%	0.0%	10.8%	4.8%	0.0%	1,984.5	$3.39
2007	7.7%	32.3%	29.3%	1.9%	0.0%	0.0%	0.0%	0.0%	20.6%	8.1%	0.0%	3,073.0	$3.50
2008	7.1%	31.0%	22.4%	3.9%	0.0%	0.7%	0.0%	0.4%	20.4%	11.5%	2.6%	5,491.8	$3.25
2009	5.2%	18.5%	15.9%	4.4%	0.0%	0.4%	0.0%	0.0%	31.9%	14.5%	9.3%	7,913.3	$2.18
2010	6.1%	14.8%	11.6%	3.1%	2.0%	0.7%	0.1%	0.0%	37.5%	16.1%	8.0%	17,402.3	$1.48
2011	3.3%	6.9%	12.1%	2.6%	2.7%	0.4%	0.2%	0.0%	46.3%	17.0%	8.6%	23,579.3	$1.37
2012	3.0%	3.7%	12.1%	2.4%	2.7%	1.1%	3.3%	0.6%	44.5%	19.0%	7.8%	26,061.8	$0.75
2013	2.1%	3.0%	10.6%	3.0%	2.3%	2.3%	2.8%	0.2%	44.6%	20.5%	8.6%	34,011.3	$0.81
2014	1.9%	2.8%	8.6%	3.0%	2.0%	1.1%	3.3%	0.0%	45.0%	23.1%	9.2%	39,397.0	$0.71
2015	2.0%	2.0%	6.4%	2.7%	1.4%	1.0%	4.3%	0.2%	48.0%	19.7%	12.4%	50,818.3	$0.72

ROW: Rest of the world.

From Fig. 4.27, it is obvious that China continues to dominate both global cell and global module assembly capacity. As the market in China was 30% of global demand, its domestic market consumes 52% of its capacity to produce modules.

Select Markets

It is useful to observe select markets whether or not these markets have proven able to successfully create sustainable market demand.

Japan

Japan is both a demand- and supply-side PV industry pioneer. Sharp Solar was the leading manufacturer of crystalline PV technology for many years, longer than any other manufacturer. Panasonic (Sanyo), Solar Frontier, (Showa Shell), and Kyocera continue as technology pioneers. In general, cell and modules from Japan are considered high quality and command a price premium in the United States and other markets.

Japan's residential rooftop incentive drove a strong market in the late 1990s and early 2000s and its recent FiT drove demand to uncontrollable heights. As the FiT-driven demand accelerated, low-cost imports from China dominated Japan's domestic market, particularly for installations into the commercial and utility-scale applications. Note that utility scale in Japan refers to PV systems >500 kWp. In 2015 with its market reaching 10 GWp, Japan did not have enough domestic cell manufacturing to fulfill its demand.

Despite high soft costs, increasing labor costs, and high costs for land, Japan learned the lesson of many other markets, that a healthy incentive will drive a market that may surge past a country's ability to control it.

Accelerated PV deployment has strained Japan's grid. Japan is an island nation and its grid is isolated and between islands there are transmission problems, something Japan has in common with the US state of Hawaii. Grid capacity constraints drove changes in the rules for Japan's FiT and its grid connection policy and stranded gigawatts of planned PV capacity. Changes included setting the FiT level on the day the grid-connection contract is signed and requiring that developers changing contracted modules to another brand also change to the FiT level in force at the time of the change.

Japan, as with other countries, is undergoing electricity market reform and unbundling utilities. Some companies are buying PV-generated electricity at higher rates than the FiT indicating the potential of a future market not driven by the FiT. Though the potential exists, it would be at a slower level of deployment.

The lesson of Fukushima has encouraged Japan's government and population to continue pursuing the deployment of renewable technologies, even with a limited return to nuclear-generated electricity. Looking forward to a future of self-consumption, the government is focused on the development of affordable battery technologies and storage is already a focus of the current deployment. Figure 4.28 depicts supply and demand market behavior for Japan from 2005 to 2015.

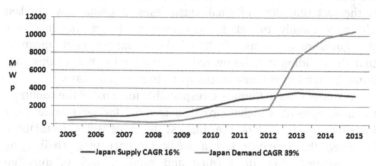

Figure 4.28 Japan supply and demand, 2005–2015.

China

China remains the global leader in PV output and will remain so for the foreseeable future. The country's manufacturers are responsible for current capacity expansions in other countries. This leadership position means that the country's manufacturers are in the position of setting the global average price and establishing module trends such as 72 cell, glass on glass, and 1,500 V. China's domestic market, the largest in the world, is almost entirely closed to outside developers with the exception of the occasional demonstration project.

China is one of the world's largest countries in terms of land mass, (9,596,960 square km land mass) with the largest population in the world, at >1.3 billion. Full, or close to full, employment is extremely important to China's central government and it has shown a willingness to support industry to continue providing jobs. Wages are increasing in the country perhaps putting a strain on its ability to continue as a low-cost manufacturing leader globally.

Though state-owned companies receive preferential loan terms, entities that are not state owned (including solar developers and manufacturers) often make use of high-interest, non-bank financing often referred to as shadow banking. These companies are vulnerable to shocks in China's economy.

China has a strong sun resource and a weak grid. Successful deployment of rooftop solar would require more and longer building ownership and currently, private ownership is new. Rooftops in China, similar to India, are not built to last 20 years. The average roof lifetime in industrialized nations is 30 years. Complicating the stimulation of a rooftop (primarily commercial) DG PV market is the fact that 80% of small businesses in China may not last 10 years. Finally, though home ownership is growing in China, the government owns 100% of the country's land, meaning that the building may be owned, but the land is not. In this way, home ownership in China is analogous to a 20-year lease.

Six provinces in China are responsible for most of the country's PV development. Qinghai province is ideal for large-scale solar installations as it has the highest level of solar irradiation; however, the province has a weak infrastructure, challenging environmental conditions (dust and sand), a lack of qualified labor, and a lack of water availability.

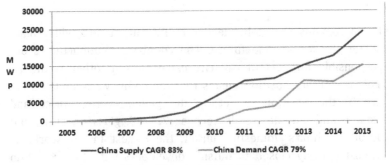

Figure 4.29 China supply and demand, 2005–2015.

The government in China is encouraging developers to offer discounts, has instituted a flexible bidding process in a move to lower its FiT rate (currently 15 euro cents/kWh) and has prioritized on-site consumption. Figure 4.29 presents the demand and supply profile for China from 2005 to 2015.

Taiwan

Taiwan continues, despite tariffs in the United States and Europe, to provide PV cells to China. As China's manufacturer expand manufacturing into other countries, less cell product will be required from Taiwan, likely driving cell prices down and pressuring margins for cell manufacturers in Taiwan. Tariffs imposed on imports by Europe and particularly by the United States have damaged revenues for the country's manufacturers.Tariffs have not reduced the demand for cells from Taiwan's manufacturers, but tariffs have forced lower prices for cells.Taiwan's manufacturers, Gintech for example, are also expanding to Malaysia and Thailand to avoid import tariffs. Figure 4.30 presents the demand and supply profile for Taiwan from 2005 to 2015.

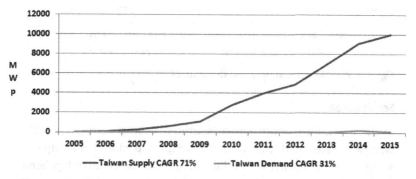

Figure 4.30   Taiwan supply and demand, 2005–2015.

South East Asia

The countries of Southeast Asia, including Vietnam, Brunei, Myanmar (Burma), Cambodia, East Timor, Indonesia, Laos, Malaysia, the Philippines, Singapore, and Thailand, have a combined population of ~600 million. These mostly island-based countries have very poor rural populations. Though the region's incentives have traditionally been spotty at best, several countries, specifically Thailand, have recently implemented programs that have elevated the region's demand profile to the accelerated scenario for at least 2 to 3 years.

Malaysia

With favorable terms, low labor costs, and other incentives, Malaysia has become a favorite country to locate manufacturing despite its unexciting market. First Solar, Jinko, JA Solar Hanwha, and others have located manufacturing in the country. Manufacturing will continue to increase for the next several years, potentially rivaling Taiwan as the second largest global cell-manufacturing countries. The risk is that Malaysia may be the next target for government import tariffs. Figure 4.31 presents the demand and supply profile for Malaysia from 2005 to 2015.

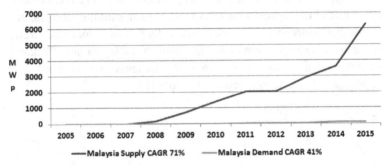

Figure 4.31 Malaysia supply and demand, 2005–2015.

South Korea

In the late 2000s, South Korea instituted a too generous FiT tariff with few rules that was oversubscribed and closed out rapidly. Since then the market for PV installations in the country has been spotty at best.

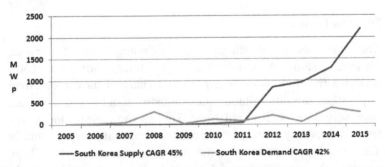

Figure 4.32 South Korea supply and demand, 2005–2015.

Manufacturers struggled to compete with low price imports from China and Taiwan and the country's manufacturers considered exiting PV manufacturing. In time though manufacturers developed a reputation for high quality and have been able to command a price premium. South Korea's manufacturers have been circumspect about adding capacity. Figure 4.32 presents the demand and supply profile for South Korea from 2005 to 2015.

India

The market for solar deployment in India has been slow to take off, but with 2 GWp installed in 2015 and an expected over 5 GWp market in 2016, its market appears to have finally gained traction. Whether or not strong growth will continue over the long term, given the country's intricate bureaucracy and low tender bidding, is up to debate.

India has a long history of PV cell and module manufacturing including participation by Tata BP Solar India Limited renamed in 2012 as Tata Power Solar Systems Limited. Currently low price imports continue to dominate its market despite government efforts to control the situation by setting domestic content requirements. Low tender bidding on projects means that domestic developers need to source the least expensive modules, typically imports from other countries. India's PV manufacturing sector currently exports cells and modules to other countries.

A large portion of India's population lives in remote rural regions of the country and rural electrification has always been an important part of government solar goals. Goals for rural electrification have not been met for a variety of reasons the most significant of which has to do with lack of sufficient funding.

Through its Solar Mission, the government of India initially established a goal of 20 GWp of multimegawatt (utility scale) solar installations as well as 3 GWp of off-grid installations. Given the country's average annual deployment of ~800 MWp annually, the realization of this goal seems unlikely. India recently increased its goal to 100 GWp of PV deployment by 2022. Figure 4.33 presents the demand and supply profile for India from 2005 to 2015.

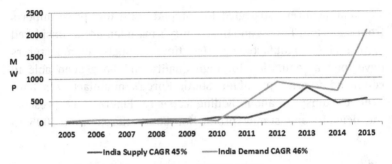

Figure 4.33 India supply and demand, 2005–2015.

The Future for PV in the Countries of Asia

The market for solar deployment in the countries of Asia is booming with three markets, Japan, China, and India, set to install ~52% of the 2016 total. This deployment will be close to 100% photovoltaic flat plate. This means that these markets are not just emerging; they have emerged and are dominating.

That PV manufacturing and deployment is not always profitable in the Asia(s) is concerning. Should the government of China pull back on its deployment goals and ease up on manufacturer support, its manufacturers would have a difficult time competing on the basis of offering the lowest price. These concerns aside, it is unlikely at this point that China's government will abandon its PV manufacturing or its domestic deployment goals.

Changes to Japan's support of its domestic market could leave it with a demand profile close to that of Europe, that is, a significant and steep decline that resembles a U-Turn in traffic.

The surging market in India could prove too expensive to support and the government could institute some of the retroactive changes to tariffs and/or PPAs that essentially destroyed the market in Spain.

China, Japan, and India are not the only growing markets in Asia. In West Asia, Pakistan is showing signs of growth, while the countries of South East Asia have growing manufacturing sectors and markets. Concerns aside, the region's manufacturers will continue to dominate PV module product for quite some time while the various country markets in the Asia(s) have room to continue growing for some time.

4.6 The Present and Future Markets in Australia and in Oceania

Paula Mints

The countries in Oceania should be a natural fit for off-grid and on-grid solar PV deployment. Yet cumulative deployment remains low. From 2005 to 2015, the region installed a cumulative 5.3 GW of solar PV, 3% of the global cumulative 211.2 GW for the same period.

The mostly-island countries in the Oceania include Australia, New Zealand, Fiji, Guam, Cook Islands, and French Polynesia, among others. Australia is the largest market for solar PV deployment in Oceania. It is also the world's sixth largest country in terms of land mass after Russia, Canada, China, the United States, and Brazil. Figure 4.34 offers solar PV deployment market shares for the regions in Oceania for 2015.

2015 Oceania 1.1-GWp

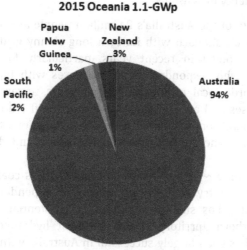

Figure 4.34 Oceania solar PV deployment 2015.

Use of solar in Oceania outside of Australia is primarily for off-grid applications, specifically, remote habitation and consumer power. Recently there have been announcements of increased solar PV deployment by some of the islands in the region to replace or compliment diesel generators. Diesel generators are the primary source of electricity for the islands of Polynesia. Tonga installed a 1.3 MW PV system (hybrid with diesel) in 2012. A resort in the Cook Islands installed an 80 kW installation in 2013.

Figure 4.35 shows application deployment for Australia and for Oceania.

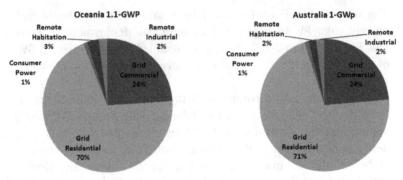

Figure 4.35   Oceania and Australia solar PV application share, 2015.

Australia, Plenty of Sunshine

The majority of the Australia's 24 million people are concentrated on the two coasts each with its own long skinny grid. Electricity rates have spiked in recent years, particularly for residential ratepayers, who responded to the increases with high levels of anger towards local utilities. In recent years, anger with utility rate increases and an overgenerous feed-in tariff led directly to a strong demand for residential solar PV installations that include storage. The generous FiT has run its course and demand has quieted.

At this time ~77% of Australia's electricity is coal generated, though the country is slowly decreasing its dependence on coal. Australia also has significant natural gas potential, particularly in its Northern Territory. A drive against hydraulic fracturing (fracking) has been largely successful in Australia with New South Wales, Victoria, the Northern Territory, and Tasmania placing a moratorium on fracking.

Australia does not have a photovoltaic cell and module manufacturing industry. The country imports the PV modules it installs. However, Australia's global influence on photovoltaic research and manufacturing cannot and should not be understated. Australia's universities (University of New South Wales and Australian University, for example) are recognized and renowned for educating today's solar industry leaders and for many of the innovations that the global solar industry takes for granted.

For example, the PERC technology (passivated emitter rear contact solar cells) currently gaining share globally, has its roots in the 1980s at the University of New South Wales (UNSW). In 1983, Dr. Martin Green and a team at UNSW announced the world's first 18% silicon cell. A few months later, the team increased efficiency to >19% with a passivated emitter solar cell (PESC). The world's first 20% silicon solar cell was produced by UNSW in 1995.

Australia's Market, Not Instant Gratification

Australia has world-class solar research but not a world leading market. Currently Australia's solar market is primarily for the residential application though the country does have a market for commercial grid-connected deployment and off-grid deployment.

The reasons for the country's low deployment of solar PV are complex though shaky government support is certainly at the top of the list. Recent government elections led to a threat to defund the country's Australian Renewable Energy Agency and though in the end funding was reduced and not eliminated, the country's government has traditionally had a love/hate relationship with renewables in general.

Australia's Renewable Energy Target (RET) is always under threat of repeal by conservative politicians. Solar, particularly residential solar, is highly popular among homeowners and thus far the RET has resisted repeal. Under the RET, Australia has a rebate or, refund, system called Small Scale Technology Certificates (STCs). Customers can choose to take their rebate several ways. They can sign their STCs over to the installer and receive a discount or direct rebate on system installation, they can apply and receive a check or bank transfer of funds from the government or they can apply and have the rebate applied to their electricity bills. The STC program is not available in all territories.

The government continues to have aggressive goals for the deployment of solar but no specific incentives to drive these goals. The country's electricity network is not designed to support bi-directional power, which is problematic for solar PV on the grid. Most of the country's population is located on one coast or the other with most of the middle of the country unsupported by grid access. Though this would seem to indicate the potential of a booming market for off-grid solar, historically this has not been the case.

Table 4.3 offers solar PV deployment history for Australia and other countries in Oceania from 2005 to 2015. The table also includes Australia's share of the global market for each year. Historically, Australia's share of global solar PV deployment has been low. In 1987 Australia had an 8% share of global PV deployment for a total market of 20 MW. In 1992, Australia's share was 5% for a total global market of 52 MW.

An Excellent Sun Resource and Unstable Policy

The sun will continue to shine over Australia and government support will likely continue to be unstable despite a stated desire to increase solar deployment. The reason for unenthusiastic government support is partly due to a focus on the short-term cost of deployment instead of the long-term benefits.

This focus on near-term costs is shortsighted as demand from homeowners in the country is strong, driven by a desire to control electricity costs and an even stronger desire by energy consumers for energy independence. Globally the solar industry has had some success inspiring people to choose energy independence. In Australia high electricity rates and distrust of utilities have done the job. Australians want solar PV systems with storage, that is, they want their piece of the sun that shines overhead.

Figure 4.36 offers a short-term forecast for solar PV deployment in Australia to 2018 from three perspectives: low, conservative, and accelerated.

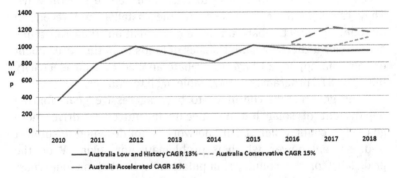

Figure 4.36 Australia solar PV deployment, 2010–2015, forecast to 2018.

Table 4.3 Australia and Oceania solar PV deployment (MW) 2005–2015

Country	2005	2006	2007	2008	2009	2010	2011	2012	2013	2014	2015	10-Year CAGR
Australia	18.8	23.6	38.9	39.7	32.7	362.9	792.3	1,001.1	904.7	817.5	1,005.4	49%
South Pacific	0.7	0.9	1.3	1.1	0.8	1.1	8.3	19.9	9.5	8.5	16.0	36%
Papua New Guinea	0.3	0.4	0.5	0.4	0.3	0.4	8.3	9.4	7.6	9.3	11.8	44%
New Zealand	1.3	1.6	2.3	2.2	1.8	1.5	16.5	14.6	30.5	11.0	35.3	39%
Total World Market	1,407.7	1,984.5	3,073.0	5,491.8	7,913.3	17,402.3	23,579.3	26,061.8	34,011.3	39,397.0	50,877.2	43%
Australia % Total World Market	1%	1%	1%	1%	0%	2%	3%	4%	3%	2%	2%	

4.7 Global Community Unites to Advance Renewable Energy: IRENA[36]

Frank P. H. Wouters

(This chapter's Glossary is on page 254)

Introduction

We live in an increasingly globalized world, a world made smaller by ever-faster transport and telecommunications and supported by global banking systems and trade rules. English has become the de facto universal language and the Internet facilitates instant communication. With the advent of globalization, whose rise started more than a century ago, the need for global governance rose. The Second World War was a huge shock to the world and after its end, the global community established the United Nations, basically to avoid another war. But it also included the World Bank and the IMF to stabilize the global financial system. Although initially the United Nations' main focus was to safeguard global peace, in the decades following its inception the United Nations established more and more agencies addressing economic and social development. There are now close to 20 specialized agencies and related organizations, including the World Bank, the Food and Agricultural Organization, the World Trade Organization, and the International Atomic Energy Agency. Although energy is the largest economic sector in the world, there is no "World Energy Organization" with a global mandate.

The period after World War II was characterized by a surge in global economic growth, which was in part supported by the United Nations. In the 1960s the period of decolonization started, and the new countries were eager to become members of the United Nations and its institutions to benefit from its supportive systems and programs. However, in the decades after the oil price shock of the 1970s, among many things characterized by the demise of the Soviet Union and the fall of the Iron Curtain in 1989, the United Nations struggled to provide adequate support to many developing countries, specifically in Africa and Asia. In the 1980s and 1990s, the international community seemed unable to effectively help developing countries in their struggle with poverty, natural

---

[36]International Renewable Energy Agency.

disasters, and armed conflicts. Most African counties had poor infrastructure and seemed unable to find pathways towards economic growth, even though many countries are well endowed with natural resources. Led by the United States, a broad call for reforms of the United Nations emerged. In a "non-paper" sent to member states just before the U.N. General Assembly opened its 50th anniversary session and entitled "Readying the United Nations for the 21st Century," the Clinton administration criticized the "unnecessary multiplication of U.N. entities" with overlapping functions, according to reports on the leaked document published in August 1995. In the years following, and in absence of wide-ranging reforms, several countries left specialized UN organizations with, for example, UNIDO losing Australia, Canada, Belgium, France, Lithuania, New Zealand, Portugal, the United Kingdom, and the United States as members. It is no surprise, therefore, that the present UN Secretary-General Ban Ki-moon, has put reforms at the top of his agenda.

There is no UN agency dedicated to energy, although on many occasions, most notably at the World Summit on Sustainable Development (WSSD), held in Johannesburg in 2002, it became clear that poverty reduction, access to energy, energy security and climate change mitigation were all interlinked issues requiring a coordinated response from the development community. This has led in 2004 to the establishment of UN-Energy, the United Nations' mechanism for inter-agency collaboration in the field of energy. However, the institutional setup is very complex, with a multitude of agencies involved as co-leads: UN DESA, UNDP, the World Bank, FAO, UNEP, UNESCO, UNIDO, and the IAEA. In 2007 UN-Energy elected Kamden K. Yumkella, then director-general of UNIDO, as chair of UN-Energy. Mr. Yumkella has since stepped down from UNIDO, as well as from his position as Special Representative of the UN secretary-general for the Sustainable Energy for All Initiative, and mid 2016 a successor had not yet been appointed.

In addition to above-mentioned UN patchwork on energy, two global agencies are worth looking into. First, there is the IAEA, which was established in 1957 to coordinate work on the peaceful use of nuclear energy and to provide safeguards against the misuse of nuclear technology and nuclear materials. Although the IAEA has its own treaty and internal governance, it reports to the United Nations General Assembly and Security Council. The IAEA currently has 168 member states and 2,300 staff members and is headquartered in Vienna. Due to its inspection and policing mandate, its annual budget, at almost $500 billion, is substantial.

The IEA was established after the oil price crisis in the 1970s by the main oil consumers, the OECD member states. IEA membership is limited to members of the OECD and over its history the IEA's focus has primarily been on fossil fuels. Nonetheless, the IEA has been an organization as close to a "World Energy Organization" as one could conceive. However, despite several decades' worth of work on renewable energy, the IEA's thought leadership position on this topic has only recently become somewhat recognized. The IEA routinely neglected the position of renewables and always underestimated the growth potential of modern renewables such as solar and wind. Lastly, since membership is limited to OECD countries, and recent growth of renewables has been largely outside of the OECD, the world needed a more inclusive organization with a truly global scope.

4.7.1 Start of IRENA

The history of IRENA goes back several decades and its roots are in Germany. The call to establish an International Renewable Energy Agency was raised for the first time in 1980 in the so-called Brandt report, which analyses the North-South divide and proposed a transfer of resources from the developed to the developing world. The establishment of such agency was consequently recommended in the final resolution of the first UN conference on renewable energy in Nairobi in 1981, the Conference on New and Renewable Sources of Energy. Despite these calls, in the decade following it was argued that it would suffice to entrust existing UN organizations with the promotion of renewable energy. Here the late visionary Hermann Scheer came into the picture, who fought tirelessly for the establishment of what he originally called an International Solar Energy Association. Scheer's organization EUROSOLAR drafted the first "Memorandum for the Establishment of an International Solar Energy Association" in 1990. The idea was widely published and presented at the

UN headquarters, resulting in the formation of a UN task force by Secretary-General Pérez de Cuéllar. The aim was to launch the agency at the first conference on environment and development in Rio de Janeiro in 1992. Between 1990 and 1992 Hermann Scheer managed to get the support of many international leaders, including Al Gore, Norwegian prime Minister Gro Harlem Brundtland, Austrian Chancellor Franz Vranitzky, former German Chancellor Willy Brandt, Indian minister for the environment Maneka Gandhi, and Rajendra Pachauri. Despite this support, the Preparatory Committee of the Rio Conference rejected this proposal. But Hermann was not a man to give up easily.[37]

4.7.2 Hermann Scheer[38]

To understand the beginning of IRENA it is important to understand the person that Hermann Scheer was.

It was one night in 1997 when I had just started my business in Cologne, Germany. A colleague of mine told me that Hermann was going to speak at the University of Bonn, a city close by. After work, we jumped on the train and went to Bonn. The event was not a big one, there were maybe several dozen students and other young people present. Hermann started slowly, his speech was not scripted, and he did not speak from paper. However, his story picked up pace and intensity, and he ended very forcefully and convincingly. He was a natural speaker, developing his story as he went. Of course he spoke about his favorite topic, the future of energy, and the role of renewable energy in our system. It was a brilliant speech, one of the best I have ever attended. It flowed logically, and it all seemed to make a lot of sense. One should remember that in 1997, renewables were a marginal piece of the energy equation, this was years before any meaningful contribution to our overall energy demand. Hermann managed to paint a very solid and convincing picture, not only from a technical point of view, but he included the political angle as well as the urgency because of climate change. And he never

---

[37]A detailed chronology from 1990 until the formation of IRENA in 2009 is described in Irm Pontenagel and Wolfgang Palz, eds. (2009). *The Long Road to IRENA*, Ponte Press Verlags GmbH, Bochum, Germany.
[38]Hermann Scheer (1944–2010).

forgot to mention the cost; he knew that renewable energy would be cost competitive in the near future. His mantra was that we should start moving away from an energy system that is based on a shrinking supply to a system based on an endless supply of renewable energy. In the first, current, situation, cost of energy will naturally go up if supply starts depleting. But if we manage to shift to renewables, supply would be naturally abundant and cost would eventually go down.

One might think that speaking for a handful of students would not change the world. However, his ability to mesmerize audiences created a group of very loyal followers. Followers that would help him achieve his goals on many levels. The organization EUROSOLAR, an organization he founded and led, was a strong advocate for the transition to a renewables-based energy system.

Hermann was also a very hard-working person. In addition to his day job as one of the longest serving members of German parliament, the Bundestag, he reached out by interacting with people in the evenings and on the weekends. He wrote many books on his favorite subjects, books that seemed utopic at the time, but that have shown pathways to where our energy system is heading now.

Since he was a political scientist and economist, not an engineer, he had a group of experts around him that gave him insights into the fields that were not naturally his. And he liked the intellectual debate. His strong views not only made him friends, there were many people that disagreed with his vision. He was not afraid to take a strong and controversial position, even in his own party. He certainly was a polarizing person people either bought into his views, or strongly opposed them. Hermann seemed to like that controversy, he drew energy from it. He also repeatedly said that the incumbent utilities would not be the main drivers of the energy transformation. There would just be too much at stake for them, they are too embedded in the fossil fuel and nuclear supply chain. He would expect them to block rather than drive the transformation. The experience of the last two decades has proven him right in this respect. In most cases, traditional utilities first ignored renewables then started to oppose them, and only after they recognized the massive and unavoidable drivers behind the growth of renewables, have they started to reluctantly embrace the new realities. If they manage to change their business model quick enough it may not be too late for some (see also Chapter 5.1). But many have seen their profits dwindle, which has been reflected in their share price, which is only a fraction of what it used to be.

Hermann continued his promotion of the idea of an International Solar Energy Association because solar energy was a universal renewable energy source, available at any place. It can be used to generate electricity and heat directly without moving parts or complicated machinery everywhere on the planet, also in developing countries and in remote areas. The other renewable energy sources—wind energy, bioenergy, hydropower, geothermal energy, and marine energy—are less universal in their application and require infrastructure, moving parts, and maintenance. However, as these renewable energy sources were also important, in a 2000 memorandum he extended the scope of the proposed organization, renaming it the "International Renewable Energy Agency" (IRENA). As we are going to see later, solar energy is still the most dominant area of IRENA's work.

4.7.3 IRENA's Roots and Early Days

An important factor behind the origin of IRENA was also the German "Energiewende," or Energy Transition, which originated from Germany's decision to exit nuclear-power, their lack of clean indigenous fossil fuels, and increasing environmental awareness of the general population. Other European countries such as Denmark and Spain were in a similar position. They all felt the need for a strong international agency that could help them find pathways for a transformation of the energy system. Unfortunately, the IEA was too much of a fossil fuel agency for them. So there was general support for the foundation of a new agency. Several academic studies have been conducted on IRENA, analyzing the mechanisms behind its foundation.[39] The fact that it is a lot more difficult and time-consuming to establish an intergovernmental organization from scratch then to refocus an existing structure, is testament to the failure of the IEA and the strong focus of the founding countries and individuals such as Hermann Scheer.

According to van der Graaf, two mechanisms can lead to the time-consuming and costly establishment of a new organization.

---

[39]Thijs Van der Graaf (2013). Fragmentation in global energy governance: explaining the creation of IRENA, *Global Environmental Politics*, **13**(3), 14–33.

First, when an international institution is captured and has lost domestic support in a group of powerful member states, the latter will attempt to create a countervailing institution. Second, when a credible countervailing institution is created, states that prefer the institutional status quo ex ante will come under pressure to join the new institution. This has happened in the case of IRENA.

As a member of the SPD executive committee, Scheer introduced the demand to create IRENA into the SPD government program for the 2002 federal elections. At the initiative of Green Party Member of Parliament Hans-Josef Fell, the German Green Party joined the SPD in this initiative. The German parliament adopted the resolution drafted by Hermann Scheer, "Initiative for the Foundation of an International Renewable Energy Agency." Due to pushback by established institutions such as the World Bank, the IEA and various UN organizations, several attempts to establish IRENA, most notably in Johannesburg (2002) and Bonn (2004) failed. Having learned from these failures, the German government changed tactics and conducted a number of bilateral talks with other governments, to create a coalition of the willing. This new approach was more successful and resulted in IRENA's first preparatory conference that took place in Berlin in April 2008 where 170 participants from 60 countries took part, discussing the possible future structure and objectives of the organization. Consequently, in January 2009, IRENA's Founding Conference was held in Bonn, with 120 national delegations from around the world attending and 75 member states signing the IRENA Statute.

Denmark, Austria, the United Arab Emirates, and, of course, Germany lobbied to host IRENA's secretariat, with Bonn, the former seat of the German government, a natural candidate. After many German ministries moved to Berlin from Bonn, the German government had been trying to attract international organizations to Bonn, and they were eager to host IRENA. However, intensive lobbying by the UAE convinced many countries, actually most developing countries, that the headquarters of IRENA should be located in Abu Dhabi. Part of the appeal was that the UAE offered to build a brand-new office building in Masdar City, the flagship project of Masdar, Abu Dhabi's future energy company. During the second session of IRENA's preparatory commission held in Sharm El Sheikh, Egypt, June 2009, Abu Dhabi was formally selected to host the interim headquarters of IRENA. At the same meeting, the French national Hélène Pelosse was elected first interim Director-General of IRENA. Since Germany had chosen to offer hosting the headquarters rather than offering the first director-general in the person of Hermann Scheer, they lost out on both fronts. However, they offered to host the IRENA Innovation and Technology Centre in Bonn, an arrangement that is still in place today. The impressive lobbying done by the UAE was accompanied by an equally impressive financial package. Compared to the US$11 million Germany offered, the UAE offered to cover US$136 million for the first 6 years. In addition, they offered US$350 million for projects in developing countries, selected by IRENA and channeled through the Abu Dhabi Fund for Development.

Hélène Pelosse had some experience in the international arena, having been involved in the European Union negotiations on energy policy. However, she was new to the world of United Nations–like organizations. Global politics are part of the fabric of intergovernmental organizations it is never "only" about renewable energy. One has to understand and respect certain things otherwise one cannot get things done. Two of the global leaders in renewable energy, China and Brazil, did not join the agency in the beginning. China took issue with the way Taiwan was initially addressed and Brazil took issue with the definition of sustainability of biomass, a strategically important economic sector for them. China joined the agency in 2014 after some of the sticking issues were addressed, and it is likely that Brazil will soon follow. Furthermore, IRENA is built on United Nations principles, which means that each country has one vote and their annual contribution is assessed based on their GDP. Tuvalu, which has less than 10,000 inhabitants has as much formal influence as the United States, but pays a fraction of the US contribution. The larger countries use their money to exert influence in an organization. The normal way is to provide voluntary contributions for activities that are of particular interest, which is something richer countries routinely do. Another way is to withhold payment, which was notoriously done by the United States in the 1980s to the United Nations. The US Congress wanted to achieve a reduction in the assessed contribution from the then 25% of the overall budget. After negotiations, the assessed contribution was indeed reduced to 22%, which has been the norm ever since. It should be noted that the United States is the only country with a fixed percentage, which is in effect a cap. Otherwise the contribution would even be higher, due to the size of the US economy. In the early days of IRENA, several countries withheld payment, causing operational difficulties for the agency. Instead of negotiating a way out, Pelosse resorted to a rather bizarre way of communicating the situation. She replaced the IRENA Web site with a note saying that the agency was bankrupt. The period in which Pelosse reigned was characterized by a number of such incidents. After a turbulent initial period, Hélène Pelosse resigned and was replaced in 2010 by Adnan Amin, a Kenyan national who had spent a career within the United Nations.

In September 2012, I was appointed deputy director-general of the agency. Although I had experience with intergovernmental agencies, my career had predominantly been in the private sector. I think that was one of the elements in my favor over other, excellent, candidates. Although intergovernmental agencies are run by and for governments, a lot of the action advancing renewables is driven by the private sector. The role of the government is to put in place favorable regulatory frameworks and policies, the private sector innovates, develops new technologies and services, creates jobs, and makes most of the investments. It is therefore important to understand how the private sector thinks, and involve them in the work of an agency like IRENA. That was part of my role. On many occasions I was struck by the fact but even people that are working on the same topic can speak a completely different language. We are all speaking English but we have a different understanding when using a certain terminology. A case in point is for example "project finance." In the private sector, it is clear that project finance refers to a loan without recourse to the investor. The future cash flows of the project serve as security for the loan, not the balance sheet of the investor. In government circles, project finance is loosely used for all kinds of ways a project can be financed. In interactions with government officials that do not have a financial background, it is important to realize that such differences in use of language exist. But I have also encountered on a number of occasions fundamental lack of understanding of what drives a business. For a business to survive, making a profit is necessary. And the profits need to be high enough to justify one type of business over another. Money flows to areas, which provide the highest return for a certain business risk. This is the same everywhere, it doesn't matter whether a business sells a TV in Belgium or a solar lantern in Uganda. However, on occasions I found that businesses catering for the rural poor in developing countries were measured differently. Some people found it immoral to make a profit off poor people. I guess part of the problem is that decades of, by the way failed, development projects, in which goods and services were handed out below market price, somehow blurred the picture. In my view, sustainability of development requires a normal profit margin along the supply chain, especially in developing countries. This is not a matter of exploiting poor people. If an element of the supply chain does not make a healthy margin, the entire supply chain breaks, and the poor people are deprived of goods and services they require. Development of rural areas can only be based on healthy business models.

Energy is the largest global economic sector and we are undergoing a rapid transformation, from a system based on limited and diminishing supply to a system based on endless renewable supply. Renewable energy is now cost competitive with conventional energy, and technical and financial innovations provide solutions that can bring modern renewable energy to the large number of people that have been deprived of development so far. Part of my portfolio as deputy director-general of IRENA was managing a project funding facility comprising $50 million per annum, over a seven-year period. IRENA in cooperation with the Abu Dhabi Fund for Development would fund 5 to 10 projects every year with the aim to provide leading examples of renewable energy applications in developing countries. Among the criteria that we defined for selection, I was always keen to see a business model that was sustainable and replicable. In that respect, one of the projects that stood out for me was a business in a non-electrified area of Mauritania that used small wind turbines to produce ice for the local fishermen to keep the fish fresh longer. The business helped build a "cold chain," which is a system of freezers from the port to the shops, thereby expanding the opportunities for the local fisherman to store and sell their catch. The link between renewable energy and increased revenue was very obvious and the sustainability aspect was not only environmental but also financial.

The following list contains some of the highlights in IRENA's young history:

- April 2011: Irena's first assembly establishes the agency with Abu Dhabi as its permanent seat.
- October 2011: Renewables Readiness Assessments Program begins.
- January 2012: High-level participation at IRENA's second assembly includes United Nations Secretary-General Ban Ki-moon and other global leaders.
- April 2012: The IRENA Renewable Energy Learning Partnership launches.
- October 2012: Construction begins on permanent IRENA headquarters in Masdar City, Abu Dhabi's clean-technology hub.
- November 2012: First International Off-Grid Renewable Energy Conference, Ghana.
- January 2013: Global Atlas for Renewable Energy, a free online resource assessment tool for policymakers and investors, goes live at Irena's third assembly.
- October 2013: Headquarters agreement between the UAE and IRENA goes into force. IRENA receives the credentials of its first permanent representatives, from the UAE and Germany.
- January 2014: At IRENA's fourth assembly, IRENA and the Abu Dhabi Fund for Development award US$41 million in loans to support the construction of renewable energy projects in six developing countries.
- September 2014: Launch of the SIDS Lighthouse Initiative and Rethinking Energy.
- November 2013: Dr. Thani Al Zeyoudi, now UAE Minister for the Environment and Climate Change, becomes the UAE's first Permanent Representative to IRENA.
- January 2015: Mr. Amin is reappointed for a second four-year term as Director-General.

4.7.4 Institutional Setup

IRENA's institutional design is a paradox. Despite the fact that the United Nations was rejected as a home for the new agency,

IRENA is designed to function in the same way as a typical UN-style intergovernmental agency. IRENA's statutes largely rely on consensus or near-consensus for decision making, which, with a near universal membership base, does not result in agility. IRENA has two main bodies: The Assembly and the Council. The Assembly comprises all its member states and is the main decision making entity. It approves the bi-annual work plan, audits the books and elects the Director-General. The Council meets twice between each annual Assembly meeting and is advising the Assembly on relevant issues. Decision-making in such a setting is inherently complex and slow. Furthermore, the first interim Director-General Hélène Pelosse was heavily criticized for taking a stance on nuclear energy and carbon capture and storage, which some vocal member states did not like. So IRENA has to carefully maneuver the waters and work on topics that are generally accepted and express opinions that are broadly acceptable. In the first few years IRENA spent most of the time and energy at the Council meetings and the Assembly hashing out organizational issues such as rotation of Council membership. This might look like a detail since IRENA's Council is an advisory body with limited real voting powers, but countries are wary of setting precedence in such a setting, which might hurt them in another organization. It was only after all of those issues were settled, which took several years that IRENA's bodies could focus on relevant renewable energy topics. Lastly, there has been broad consensus that IRENA should not be another implementing agency, but rather a think tank or change agent. There are many examples of UN agencies that have small country offices in many countries, with resulting large overhead costs and arguably low impact on the ground.

4.7.5 Hub, Voice, Resource

IRENA has three main functions. It intends to be the global hub for renewable energy, the unified voice for renewable energy stakeholders and the resource, where all relevant information can be found. The hub function is expressed in the three main annual meetings, two Council and one Assembly meeting where close to 100 countries are usually present. The resource function is carried out by all the publications IRENA produces, but also in the form of, for example, the Costing Alliance, which provides detailed information on real costs of renewable energy around the globe, and products such as the Global Atlas, a global online database of renewable energy resource. The voice function is still work in progress. Traditional media outlets routinely refer to more established sources such as the IEA and Bloomberg, even for renewable energy information, because they have delivered regular data using proven channels for much longer than IRENA. However, after several years of producing data on jobs in renewable energy for example, one slowly sees that the media is starting to recognize IRENA as an authority. IRENA's publications are all available on the Web site[40] at no cost. Also, since IRENA's mandate focuses on renewable energy only, there is a natural tendency for media to label IRENA an advocacy group. The challenge for IRENA is to find a place in the global energy debate, not just renewable energy. After COP21 in Paris and the continuing strong growth of renewable energy all over the world, I think it is just a matter of time before IRENA's excellent publications find their way into mainstream media outlets.

4.7.6   IRENA's work

IRENA works on all six forms of renewable energy: solar energy, wind energy, bioenergy, hydropower, geothermal energy, and marine energy. With the exception of geothermal energy, which is fed by nuclear processes in the earth's core and tidal energy, which is caused by the moon, all other forms originate from the sun. The sun makes plants grow, makes water evaporate, and hence stimulates the water cycle and causes areas of high and low pressure, causing wind. More directly, solar energy can be converted for direct heat use or for the generation of electricity using a thermal cycle or solar cells. Solar energy, in particular solar PV, is the most universal of all renewable energy types, because it can be used everywhere, even indoors. Applications range from tiny cells powering calculators and watches to large ground based power plants. In fact, there is no other energy type as versatile and universal as solar energy. When conducting a search on IRENA's Web site early 2016, this fact was made obvious by the large number of references found for "solar energy," i.e., 1,910.

---

[40]http://www.irena.org/Publications.

This compares to 1,063 for wind energy, 860 for bioenergy, 719 for hydropower, 605 for geothermal energy, and 465 for ocean energy.

In its shorts history, IRENA has worked on a number of topics that stand out and that are making an impact.

The Jobs Reports

Every year IRENA produces a report that shows the number of jobs in renewable energy, both direct and indirect, on a global scale. This was something that did not exist before. Of course, people knew that jobs are created in the sector, but this was never quantified on a global scale. Lobbyists for other sectors even argued that due to the loss in other sectors, the net benefit may be negative. It was, and is, important to have those numbers and build a case for renewables as an economic growth engine for the future. This report quickly became the global standard for the topic and is widely used by every organization reporting on renewable energy.

Following the proven and trusted methodology, IRENA now routinely produces reports on jobs in countries and regions, in the framework of other reports.

Renewable Readiness Assessments

Governments routinely ask IRENA for advice on policies measures for their countries. As a first step in that process, IRENA has developed the so-called renewable readiness assessment or RRA approach. The innovation does not lie in the academic soundness of the analysis, because any experienced consultant can fly into a country, assess the situation, assess the potential, recommend actions, write a report, and fly home again. What is different with the RRA is the process, and the involvement of all stakeholders. Very often the policy mandate to advance renewable energy is not vested in the most powerful ministry. In many countries it is for example in the ministry of the environment, and not in the energy ministry or the ministry of economic affairs, where all the main energy related decisions are taken. The RRA approach tries to provide a platform where all the relevant actors come together, are involved, and hence take ownership going forward.

An RRA is usually a first step, after which a renewable energy target is formulated, a renewable energy auction is organized or another policy framework is implemented.

Renewable Energy Roadmap (REMAP)

In 2012 the secretary-general of the United Nations started an initiative called Sustainable Energy for All, in short SE4All, which has three elements. By 2030, the initiative aims to double the share of modern renewable energy, double the rate of energy efficiency, and provide universal access to modern energy for everyone. The baseline is 2010. IRENA has been selected the lead agency for the renewable energy doubling target. The backbone of IRENA's work has been the REMAP (Renewable Energy Roadmap) exercise, which analyses options to reach the doubling target. The chosen approach has been to start with 25 countries, representing 75% of global energy demand, analyses their targets and policies and reflect on them in close consultation with local stakeholders and influencers. In nearly all cases, national policies and targets would not lead to doubling the share of renewables, but would fall substantially short of that. The power of REMAP's bottom-up approach is that the options are not selected in a vacuum but are constructed based on national priorities and approaches, in close consultation with national experts. In many cases they are simply more ambitious or stretched policies and targets then already present and debated in a national context. Furthermore, the REMAP team provides the national stakeholders with an excel sheet used to calculate the options leading to the doubling target. This level of transparency has proven instrumental in achieving an acceptance for and buy-in of the approach.

Lastly, the REMAP options provide pathways for a sound economic growth, while staying below a global warming level of 2°C, which is the level below which catastrophic climate change might be contained.

Clean Energy Corridors

One of IRENA's strengths is its convening power. Routinely delegates from close to 100 nations convene in Abu Dhabi, several times per year. This enables debate and coordinated action on a national, regional and global level. There are many synergies to be obtained in linking regional energy markets, specifically electricity markets. Many countries are connected and organized through power pools. In Africa, for example, there are several regional power pools, with functioning wholesale markets. Expanding renewable capacity in such markets can benefit from cross-border sales, leading to substantial savings and a higher overall level of supply security. The Africa Clean Energy Corridor initiative was first introduced at the third IRENA Assembly in January 2013. IRENA organized a two-day workshop on June 22–23, 2013, in Abu Dhabi, which brought together major regional bodies, power pools, utilities, independent power producers, ministries, financial institutions and development partners to discuss what could be done to increase the share of renewables in future generation plans. The Initiative was consequently endorsed by ministers from countries of the Eastern Africa Power Pool (EAPP) and the Southern African Power Pool (SAPP) at the fourth IRENA Assembly in January 2014. The initiative calls for accelerated deployment and cross-border trade of renewable power in a continuous network from Egypt to South Africa.

Part of the problem with regional energy planning is that often the data that lies at the core of an energy plan or strategy evolves very dynamically, which is especially the case for renewable energy. The efficiency and cost for coal-fired power or hydropower has not changed very dramatically over the last couple of decades, so it is easy to make a forward-looking planning. Solar and wind power, however, now only cost a fraction of what they cost even a few years ago. There are not many countries that have taken a correct forward-looking view of the cost of solar and wind power in their energy planning. A regional plan is typically built up using national plans. Considering that national plans have to be approved by the government and sometimes even Parliament, one can understand that a regional plan is always based on data that is several years old. I have not seen a plan that gives sufficient credit to the current cost and efficiency of modern solar and wind; they are always under-represented.

The Africa Clean Energy Corridor intends to help by providing tools and support to arrive at a more accurate picture for Africa's future energy system, incorporating much more modern renewables then before. The addition of more cost-effective renewables will save cost, reduce price volatility and increase security of supply. The Africa Clean Energy Corridor initiative has a number of elements, ranging from planning, training, access to finance to support on technical aspects of modern renewables.

Similar corridors are under way in the MENA region, Asia, and Latin America.

Islands are special for IRENA. From a technical and economic point of view, islands are leading the pack when it comes to transforming the energy system from one based on fossil fuels to one based on renewable energy. Most developing islands, which are categorized as small island developing states (SIDS), import diesel to run a generator. Although there has been a slight relief since 2014, the high oil prices in the decade before that put fiscal strain on the countries' economies. The increasing cost competitiveness of modern renewable energy technologies is opening opportunities to replace diesel generated electricity. The challenge is that, quickly, such a system approaches very high penetration rates of variable renewable energy. Since most islands are not connected to other grids, they need to manage this variability by integrating battery storage, by hybridizing the grid or applying other smart grid technologies.

IRENA has done a lot of work with islands in the Pacific Ocean, the Caribbean and elsewhere. The first approach is typically to take stock, assess the potential for renewable energy, and have a good look at the grid. Most tropical islands have sunshine, so photovoltaics should always be an element. However, many islands do not have a lot of spare space so there is a limit as to how much solar can contribute. Wind is often complimentary, but many islands in the Pacific are subject to hurricanes, so special smaller wind turbines are used that can be lowered in case a hurricane is approaching. Some islands have good potential for geothermal energy, and some could use biomass or waste to produce electricity. Larger islands often have hydropower potential.

Another reason why IRENA focuses on islands is that they constitute a substantial part of the membership. In a one-country one-vote system, that is an important consideration. IRENA is and remains of course a political institution.

4.7.7 The Way Forward

IRENA has grown tremendously fast since its inception and can boast about an almost universal membership. As of May 2016, IRENA had 147 Members and 29 States had started the formal process of becoming Members. Especially in view of the fact that many other UN organizations are struggling, IRENA's success is striking. Also, IRENA managed to increase its annual budget in the early years, while many other organizations had to manage their operations with shrinking or zero nominal growth budgets. The international community clearly thinks a global institution working on renewables is necessary and useful. A number of products IRENA produces stand out and the agency has, for example, become the de facto authority on renewable energy jobs data.

However, its current structure, which is based on UN-style templates and consensus, is heavily politicized and the slow decision-making process does not provide for much agility. To avoid catastrophic climate change, the global energy transformation needs to accelerate and renewable energy plays a crucial role in that. However, there are not only winners and there is and will be opposition to change. The future will tell whether IRENA is in the right position to contribute beyond providing nice reports and data about the benefits of renewables. The agency will need to take positions in the overall energy debate in an actual sense and in the spirit of its founder Hermann Scheer. If not, it will remain a center of excellence, providing background information, and others will become more relevant in the transformation process.

4.7.8 Glossary

ARRA	:	American Recovery and Reinvestment Act of 2009
FAO	:	Food and Agriculture Organization of the United Nations
IAEA	:	International Atomic Energy Agency
IEA	:	International Energy Agency
IRENA	:	International Renewable Energy Agency
MENA	:	Middle East North Africa
OECD	:	Organization for Economic Co-operation and Development
REMAP	:	Renewable Energy Roadmap
RRA	:	Renewable Readiness Assessment
SPD	:	Social Democratic Party of Germany (German: Sozialdemokratische Partei Deutschlands)
UAE	:	United Arab Emirates
UN DESA	:	United Nations Department of Economic and Social Affairs
UNDP	:	United Nations Development Program
UN-Energy	:	United Nations Energy
UNEP	:	United Nations Environment Program
UNESCO	:	United Nations Educational, Scientific and Cultural Organization
UNIDO	:	United Nations Industrial Development Organization
WB	:	World Bank
WSSD	:	World Summit on Sustainable Development
WTO	:	World Trade Organization

5

The Impact of Solar Electricity

5.1 The Impact of Solar Electricity

At the beginning when PV was expensive, solar electricity for terrestrial purposes was mostly used where utilities did not exist or where it would have been too expensive to extend power lines to where the need was. However, even in those days, the utilization of PV had a large effect on other sources of electricity and on certain industries. From the many examples, below are a few interesting one:

Navigational aids. The utilization of PV to charge batteries revolutionized the maintenance cost and operation of navigational aids, lighthouses, and buoys. From the turn of the 20th century, large rechargeable batteries and diesel generators were used to power lights and foghorns in navigational aids. One can imagine the expense to supply diesel fuel to extremely remote lighthouses or periodically exchange batteries in thousands of buoys, many of them located in remote waters. Navigational aids were an ideal, important, and instantaneous application and market for solar electricity. At present there are about 13,000 lighthouses and tens of thousands of navigational aids (the US Coast Guard has probably at least 5,000). The savings utilizing PV to charge the batteries in navigational aids with PV for the Canadian and the US Coast Guards were substantial. President Ronald Reagan commended Captain Lloyd Lomer[1] of the US Coast Guard for "saving a substantial amount of the taxpayer's money through your initiative and managerial effectiveness as project manager for the conversion of aids to navigation from battery to solar photovoltaic power." There are also a large number of structures on the water, most of them related to oil, which have to be marked. By now, practically all navigational aids are solar powered.

Kerosene. Another of PV's effects was a reduction in the use of kerosene. Kerosene is used for lighting in a big part of South Asia and Africa. Climate and Clean Air Coalition (CCAC) estimates

---

[1]President Ronald Reagan's letter to Captain Lloyd R. Lomer dated September 29, 1986.

Sun towards High Noon: Solar Power Transforming Our Energy Future
Peter F. Varadi
Copyright © 2017 Peter F. Varadi
ISBN 978-981-4774-17-8 (Paperback), 978-1-315-19657-2 (eBook)
www.panstanford.com

that globally 250 to 500 million households, about 1.3 billion people, use many thousands of metric tons of kerosene for illumination. The hazard of using kerosene is twofold. It is dangerous for the people who use it in lamps because several times, kerosene lights ignite their surroundings, resulting in the destruction of property and injury or even death. Burning kerosene for illumination is also very bad for the environment.

In Africa, about 60% of the population has no access to electricity and depends on kerosene for lighting. The World Bank Group (WBG) started a "Lighting Africa" program in 2007 promoting solar lanterns. The program was started in Nigeria and continued in Kenya and Ghana[2] and it was extended to other parts of Africa. The WBG developed standards for durable solar lanterns to be used for individuals and families useful also for the education of children as well as for small businesses. These solar lanterns have now also been made capable of recharging mobile phones, as described in Chapter 4.2.1.

India has 400 million people (about 80 million households) using kerosene lamps. In India, where most of the WBG-specified solar lanterns are now built, the government is supporting the program to replace the kerosene lamps with solar lanterns.

As a result of the utilization of solar electricity, very soon a large amount of kerosene will not be used for lighting anymore.

Water pumps. PV also had an interesting effect on the pump industry. The first solar-powered water pump was implemented by Wolfgang Palz of the European Commission in 1974 on the island of Corsica, France.

In Africa, outside large cities the electrical distribution is practically nonexistent, but drinking water is needed for the large population, as well as for animals and for agricultural irrigation. The importance of water can be seen from the fact that 39% of the inhabitants of the Sahel region[3] do not have access to safe drinking water. At the same time, under huge and extremely arid areas such as the Sahara, covering over 9,400,000 km^2 (3,600,000 square miles), and south of it the Sahel, covering an area of 3,053,200 km^2 (1,178,800 square miles), are several major aquifers

---

[2]WBG Lighting Africa—2015 News Archives.
[3]Sahel forms a 5,400 km (3,400 miles)-long belt south of the Sahara, spanning Africa from the Atlantic Ocean in the west to the Red Sea in the east, and varies from several hundred to 1,000 km (620 miles) in width.

that hold an incredible amount of water in pores of stones, like a wet sponge.

One of these, under the Sahara, extends over 2 million square kilometers and underlies Egypt west of the Nile, the western part of Libya and Sudan. This aquifer has about 400,000 km³ of water—the equivalent of 4000 years of Nile River flow. It is called the Nubian Sandstone Aquifer System. In this system, the water table is at most 100 m (330 ft.) and in some areas as close as 2 m (7 ft.) below the surface. It is practically independent of the amount of rain over this huge area, and its quantity seems to be rising and not dropping. This means that pumping water by PV would be very feasible. The European Union (EU) implemented a Regional Solar Program (RSP) in the Sahel by installing a large number of solar water pumping systems. The EU's RSP-1 program was carried out from 1990 to 1998 and produced excellent results. It was continued by the RSP-2 program completed in 2009 and provided more than a million people with drinking water. Under RSP-2, 1,000 solar water pumps were installed, as well as 16 pumping systems for irrigation. These projects were also extremely successful.

Since that time, the price of solar modules has dropped fantastically. Encouraged by the RSP program's success, more and more solar water pumps for drinking water and also for other applications, even for swimming pools, were installed in Africa, as well as in India and in many other parts of the world. The effect of this was that electric pump manufacturers started to design pumps specifically applicable for solar water pumping. The demand became very high and as a result, in 1996 one of the large electric pump manufacturers, Bernt Lorentz GmbH, Germany, decided to exclusively manufacture water pumps for solar applications. The company has representatives now in over 100 countries. Today one can find a solar water pump dealer in most of the larger cities in Africa as well as in South India.

These are a few examples, but many more could be described where PV changed the existing industry, or the utilization of existing products, or a new industry was started. In the following two chapters, the case histories of PV's interaction and effect on two major and important industries, the oil and electric utilities, and as a result the future of these industries will be discussed.

5.2 In the Twilight of Big Oil, in Retrospect, PV Was a Missed Boat

Why should oil companies get involved in photovoltaic power generation? Their point of view was that if the sun's nuclear energy was going to replace oil or gas, the fuel they were selling, then they should sell that, too. By the end of 1983, the three major PV manufacturers in the world were owned by oil companies. It is, however, interesting to look at the entire picture. Which oil companies got involved in PV, and what were they doing?

Solarex was started in 1973 as a privately financed PV company and dominated the PV business for a decade. **AMOCO** at some point invested in Solarex. Solarex was a rapidly expanding company and needed lots of cash for its expanding operation. To get listed on the stock market and raise money from the public was impossible. In 1983 when Solarex was the world's largest manufacturer of solar electric products, AMOCO acquired it.

Exxon started Solar Power Corporation (SPC) also in 1973 and concentrated mostly on expanding their government PV project businesses. After President Reagan slashed the PV budget, SPC's excellent technical people tried to sell more to the commercial market, but commercial sales have to be built up and that cannot be done very fast. Losing the government demo business and not being able to replace it with commercial, Exxon discontinued Solar Power Corporation in 1984 and sold its assets to Solarex.

Arco acquired STI in 1978—it was started and founded by Bill Yerkes in 1975—and renamed it Arco Solar. Arco Solar had good technical staff as well as good management and became the largest PV firm after 1983, but in spite of that ARCO sold Arco Solar to Siemens in 1989.

Shell first started to invest in Solar Energy Systems (SES) manufacturing thin film CdS/Cu_2S solar cells in 1973. Later Shell invested more, up to approximately a total of $80 million.[4] SES was shut down because of the degradation of the CdS/Cu_2S solar cells in the terrestrial environment. Shell subsequently went in and out of several PV businesses. I research the stories I am writing very thoroughly, but I would like to say that I would

---

[4]Karl W. Böer (2010). *The Life of the Solar Pioneer*, iUniverse, Inc., p. 221.

not guarantee the accuracy of the description of Shell's zigzag in PV. The end result was that the only PV business Shell presently is involved is in Japan, Showa Shell Solar, which recently was renamed Solar Frontier and is producing large quantities of copper indium gallium (di) selenide (CIGS) thin film modules.

Mobil was involved with Tyco in the development of the EFG (Edge defined Film-fed Growth) silicon sheets, but Mobil-Tyco Solar Energy Company did not continue after the EFG process was developed. EFG-based solar cells were manufactured by Schott, German, but as it turned out, it was not competitive with crystalline silicon wafers made by other methods. Schott discontinued its production by 2012.

BP's PV division was started in 1979 and subsequently in 1981 it acquired the solar part of the British Lucas Energy Systems. It was renamed BP Solar. As a result of BP's merger with AMOCO, Solarex and BP Solar were merged under the name of BP Solarex (several years later, it was renamed BP Solar) with its headquarters in Frederick, Maryland, where the Solarex headquarters were. In 2011, in the middle of the large demand for PV modules—and despite that all of BP Solar's production was sold out and they also had a large backlog of orders—BP closed all of its PV operations.

More recently, in 2015, several articles were published that realized and explained the changes in the oil and gas market and its effect on the oil industry. For most of the past 40 years, OPEC, the association of Big Oil exporters, and the big international oil companies controlled our lives, but they have started on an inevitable decline. Competition from renewables and smaller players, as well as tighter climate polices, will make their business model obsolete. Their corporate culture makes it unlikely they will be able to adapt. The destructive effect of their "corporate culture" is to prohibit the companies from moving into a new but related field.

OPEC was initiated in 1960 by five countries and by the 1970s had 12 members who controlled the flow of oil and its price. Everybody remembers the 1973 oil crisis, which ushered in a recession in the USA and other countries. In the following years OPEC regularly made threats and on at least one occasion caused another crisis, in 1979. They forgot the age-old saying that one should not make war with one's customers. The customers will react somehow.

And they did—as is evident from the fact that today among the world's top 10 oil-producing countries, only four are members of OPEC. The oil production in the USA, which was the hardest hit by the 1973 "oil embargo," now exceeds that in legendary oil producer Saudi Arabia by no less than 12.5% and the Russian Federation by 29%. Among the 20 oil-producing countries with daily production over 1,000,000 bbl/day (barrels per day), there are only 10 OPEC countries.

A case in point is Brazil, where crude oil production in 1980 was 182,000 bbl/day and in 2014 catapulted to 2,950,000 bbl/day—more than what is produced in the OPEC countries Kuwait, Venezuela, and Nigeria.

OPEC is at a point where it is falling apart. Now it consists of a group of haves and have-nots. The have-nots want the largest, Saudi Arabia, to decrease production, but they do not understand that we are not in the good old days anymore. There is now plenty of oil and the new exploration method, fracking, has turned the tables around.

The conclusion is as BP's Chief Economist Spencer Dale said recently, "OPEC is simply powerless." If this would be the script of a movie about OPEC, the next frame would have only two words: "THE END."

The Big Oils, international oil companies ExxonMobil, Shell, Chevron, BP, Total, and others, have also gone past their zenith and started on their twilight journey. In the future, they may either transform themselves (as some German utility companies are attempting to do) or become immaterial, as happened to companies such as Kodak, RCA, Xerox, Polaroid, and many others—names the present generation does not even remember. The similarity between these forgotten companies is that they all rode up to the top and then went down and became obsolete. The reason was not that people stopped making photographs or buying TV sets or making copies of their papers—the reason was that they started doing these things differently, and the companies' "corporate culture" made it impossible for them to adapt. Thus, for example, Kodak's management and technical staff could not believe that digital imaging could threaten the traditional film business. Their "corporate culture" did not let them do it, and 124 years after it was founded, it filed for bankruptcy in 2012. There are many similar examples to show that the destructive effect of the "corporate culture" is to prohibit companies from moving into a new but related field, which ultimately could make their dominant business secondary or even obsolete. And that is happening now with the Big Oil.

John D. Rockefeller's oil empire was structured to control the oil refinery and distribution business. ExxonMobil, a descendant of Rockefeller's Standard Oil, is still the largest refiner in the world. It is hard to pinpoint the time when oil companies started to concentrate on finding and producing more and more oil and gas. But by the 1960s, they had developed expertise and money to carry out very large projects, which came to be the hallmark of the "corporate culture" of Big Oil.

Drilling to find oil started 156 years ago in 1859 in Pennsylvania with the 69.5-foot (21.2 m)-deep "Drake well." As demand increased the oil companies developed technology to find new oil and gas reservoirs. They had to go deeper and deeper to discover new oil formations, many of them offshore under deep sea water, and lately even in the Arctic. Today a great number of drilling rigs are being used, which can operate in water deeper than one mile (1,600 m) and can drill 5 miles (9,000 m) or deeper. In 2012 ExxonMobil completed the world's deepest well on the Sakhalin shelf in the Russian Far East: 7.7 miles (12,376 m) deep and 7.1 miles (11,426 m) out under the ocean.

The cost of these deep and offshore drillings is unbelievably high. The daily rental cost of a deep water drilling rig used, for example, in the Gulf of Mexico is about $500,000 excluding other expenses. Because of their expertise and wealth, this type of drilling assured Big Oil a dominant position. The cost of drilling was immaterial because the upward elasticity of the price of oil seemed to be infinite.

The price of oil (per barrel) at the beginning of 2000 was $25. This lasted for 3 years when in 2003 it started to move higher. In 2005, it had doubled and 3 years later in 2008 doubled again and reached $100. From 2008 the price of oil fluctuated between $90 and $110. However, by the middle of 2014, the price started to decline sharply and by the beginning of 2015 it was $50—half of what it was 6 months before. Since then it has been in the $30 to $50 range, which was the price of a barrel of oil 11 years ago (data: US Energy Information Administration).

The stability of the price of oil from 2008 to mid-2014 prevailed in spite of the increased consumption in China and India and of interruptions from major suppliers such as Libya, Iran, Iraq, and Venezuela. These were counterbalanced by improved efficiency of products using oil, such as cars, switching from oil to natural gas in electric power stations and the beginning of the tightening of policies related to global warming. The high price of oil also encouraged more drilling, and currently more than 100 countries are producing more than 1,000 bbl/day.

Further, there was another reason why prices eventually came down. A new technology called "fracking" to extract oil and gas from shale started to be used on a large scale. Drilling for oil deeper and deeper at very challenging locations became extremely expensive in comparison. It required enormous amounts of capital, and therefore, it was a small club that was able to do it. Fracking required little money, and therefore lots of startups got in.

Fracking experiments, the injection of fluid into shale beds at high pressure in order to free up petroleum resources (such as oil or natural gas), were started in the 1950s. Large-scale utilization first occurred in 1968 to improve the production of vertically drilled oil or gas wells. When it was realized, in the 1980s, that oil and gas inclusions in shale were in many cases horizontal and not vertical, horizontal "fracking" to create oil and gas wells was started. But more generally, it was used from 1991 onward.

Nonetheless, in 2010 still only a negligible amount of oil was produced in the USA by fracking. Five years later, in the beginning of 2015, close to 50% of the US production was from fracking. To appreciate how much oil is produced in the USA by "fracking," one should consider that at the beginning of 2015, daily production was somewhat more than the production of two OPEC countries, Kuwait and Algeria, put together.

Big Oil knew about this and could have easily branched into fracking when it was still in its infancy. But Big Oil's "corporate culture" would not let them to believe that their well-established and successful drilling technology could be affected by shale fracking. As Big Oil ignored it, a number of small organizations were able to get started.

Shell, for example, ignored fracking for a long time. When they did get in, they were, as Karel Beckman[5] writes in his recent article, "simply unable to survive in this kind of highly competitive market in which small, versatile players set the tone." Please remember that Kodak also entered the digital camera business but had to close it because, similarly, they could not compete with the many relatively small organizations that had entered that field.

The story of Shell's "Polar Pioneer," the oil rig used to start the exploitation of the very large oil resources in the Arctic, will be a famous milestone in the history of Big Oil. Until the end of the summer of 2015, Shell's management seemed to still believe in Big Oil's motto: "Damn the cost of drilling and full steam ahead." They towed the gigantic oil rig to the Chukchi Sea, offshore Alaska, and paid $620,000 per day during the summer drilling season and $589,000 a day for the rest of the year, to lease the rig.

"Polar Pioneer" started drilling on July 30, 2015. But Shell obviously realized that under the new market circumstances, the expensive exploitation of Arctic oil would not be profitable and that this condition could last for a decade or more. On September 27, the company announced that it would pull back from oil exploration in Alaska and started to tow the Polar Pioneer back to Seattle. We can mark this date, September 27, 2015, as the day the twilight of Big Oil's dominance started.

The question now is, will Big Oil find new areas to grow?

As Shell demonstrated, the "corporate culture" of Big Oil makes it unlikely that it will be able to adapt to the world of "fracking." Another problem is described in the recent book *The Price of Oil*, by R. F. Aguilera and M. Radetzki. The world is headed for an era of oil "superabundance" in which the low price of oil will prevail and oil produced with oil rigs costing $500,000 per day can only be sold at a loss. This would mean not only loss of profit but also lower revenues.

The oil companies could reverse this trend and diversify into the field of renewable energy. But this will be difficult for them to do. I know this from my own experience. As I describe in my

---

[5]Karel Beckman (2015). Exit ahead: Shell at the end of the oil superhighway. *EnergyPost*, September 28.

recent book, *Sun above the Horizon*, around 1973 the oil companies got involved in the solar photovoltaic business. There were two reasons for this:

(1) The fashionable doomsday reason: Oil and gas will run out, solar energy is permanent. Oil companies should get in now (in the 1970s) to invest in the development of the continuation of their oil and gas business.

(2) The other reason was that a few leaders of the oil industry correctly envisioned that PV, because of its decentralized nature, would become an independent energy source parallel to oil and gas.

As described before, the terrestrial PV industry was started in 1973 but by the end of 1983 the major PV manufacturers in the world were all owned by oil companies, but ultimately all of them got out of the PV business. Why? The major reason is their "corporate culture."

So for Big Oil to get back now to PV is the same as getting into "fracking"; it is the "corporate culture." In the days when Big Oil calculated their odds to invest in PV, their only risk was that they would lose their investment, which was not much more than the loss of a single dry hole. But even the minimum investment today for Big Oil is much bigger. Two big oil companies are now in the PV business but both are only minor participants.

As mentioned before, Shell's Showa Shell Solar (Japan) was started in 2006, manufacturing the thin film CIGS solar cell/module. The company was renamed Solar Frontier in April 2010. Solar Frontier's manufacturing capacity will reach over 1 GW in 2015, which is about 3% of today's PV manufacturing. The French Total Oil invested 1.1% of their entire market capitalization to buy 60% of SunPower (a US PV manufacturer). Today (October 2016), if Occidental Petroleum (OXY) were to buy 60% of SunPower (SPWR), Occidental would have to put up about 1.2% of its entire market capitalization. If OXY were to buy 60% of SolarCity (SCTY), it would need to put up 2.1% of its market capitalization. SunPower and SolarCity are the relatively small businesses in the PV industry.

The Paris climate agreement makes it clear that the world will not turn back on serious global warming policies. The oil industry started to feel the impact of fracking and also the renewable energy, specifically solar and wind. There do not appear to be many options left for Big Oil. If they want to avoid the fate of Kodak, they could split up like German utilities E.ON and RWE did, and shed their drilling business, in the same way telephone companies did with their landline business. After that in their twilight, they are not going to control our lives; they will have to compete like every other industry is doing.

The PV industry should be grateful to the Big Oil

Reading this chapter, the reader may conclude that the writer is one who believes, like many other people, that the oil industry was the Big Bad Wolfe for PV. On the contrary, I was the cofounder of Solarex and was personally knowledgeable of Big Oil's involvement in PV. Based on my experience, I can state that the oil companies provided money, which no other investor did, for the establishment of the first terrestrial solar cell and module factories.

Examples are Shell (Solar Energy Systems), Arco Solar, Mobil Solar, Exxon (Solar Power Corporation), AMOCO (Solarex), BP (BP Solar and Tata-BP), and TOTAL in France. Without the Big Oil money, these companies would not have been able to develop the terrestrial solar cell and module production and establish their manufacturing facilities, which is the basis of the PV industry of today. Without the Big Oil money, the terrestrial PV manufacturing process and market would not have been established.

5.3   PV and the Brave New World of the Electric Utilities

"The energy market of the future is green, decentralized and digital."[6]

Global warming is the reason that renewable energy (RE)—wind, bio, geo, and solar (PV) is getting lots of publicity. These renewables are now important because the electricity they produce can be used to replace coal and fossil fuels, which are leading causes of global warming.

There is, however, a big difference between the electricity production by wind, bio, hydro and geo compared to the electricity produced by PV. Namely wind, bio, hydro and geo systems produce electricity at a central location where their "fuel" such as wind, water, biomaterial, or geo heat, is available. This could be anywhere, mostly in a remote location far from where the electricity is being used. For this reason, the electricity produced by these sources need the electric grid, the transmission lines utilized by the traditional generators using coal, fossil fuel, or nuclear.

PV is, however, different. PV systems can be deployed in a variety of ways. They could be deployed everywhere as a centralized power station in hundreds of megawatts or gigawatts size like coal, fossil fuel, nuclear, and other RE power-generating systems and connected to the grid. The difference is that PV can also be installed in small watt, kilowatt, or megawatt size electricity producing systems anywhere where electricity is needed without connection to the grid. This means that the small and large PV systems could be connected to the grid and/or could be used directly connected to the user. This flexibility is one of the reasons that PV is able to revolutionize the 100-year-old terrestrial electricity-producing industry.

The utilities believed their centralized electricity business is secure because they had insurmountable barriers protecting their monopoly. One of the barriers was that independent power producers (IPP) could sell their electricity only if they could reach the customer, and for this they had to connect their generating

---

[6]Peter Terium, CEO of the German RWE, one of the world's largest electric utilities, Press Release about "innogy": RWE International SE June 29, 2016.

system to the existing grid. Obviously, the utilities vehemently objected to PV or any other RE sources of electricity connecting to the grid.

As described in Chapter 2.1, this barrier was removed starting with the "Public Utility Regulatory Policies Act" (PURPA) passed in November 1978 by the United States Congress as part of the National Energy Act and in 2003 when the Institute of Electrical and Electronics Engineers (IEEE) standard 1547 "Standard for Distributed Resources Interconnected with Electric Power Systems" was introduced.

The utilities still felt safe that the high cost of a power plant utilizing coal, gas, or nuclear fuel was also an insurmountable barrier to prevent competition from being established.

Table 5.1 provides information about the capital cost[7] of power plants operating with various types of fuels. All the information in this table—except where noted—is from the United States Energy Information Administration (EIA).[8] The capital cost for each of the systems based on their size and how much electricity they generate is provided in US dollar/kW.

Table 5.1 indicates that the least expensive electric power station is the one utilizing natural gas, even if it is equipped with a carbon capture and sequestration (CCS) system. Obviously, an additional expense is that it requires a gas pipeline to be built. Furthermore, the gas price is variable. It is also noticeable, that the EIA list does not include any power plant utilizing oil as its fuel. The most expensive systems are the biomass, offshore wind, and nuclear (uranium).

The capital cost of a PV utility-scale system in 2012 when these data originated is in the range of the coal-powered power stations for which an additional expense is a railroad line and a proper road system. PV does not require either of those. There are, however, reasons to believe that today the cost for the PV systems is even lower. The EIA report indicated that the PV system cost in 2012 was 20% lower than in 2010. The PV module prices during that time period decreased about 34%. The PV module prices in 2015 were 30% lower than they were in 2012; therefore, another 20% decrease for the PV systems can be considered.

---

[7]Cost is in October 2012 US dollar.
[8]US Energy Information Administration (April 2013). Updated Capital Cost Estimates for Utility Scale Electricity Generating Plants. http://www.eia.gov/forecasts/capitalcost/pdf/updated_capcost.pdf.

Another issue is that the EIA report states that the capital cost ($/kW) of the 150 MW PV system in Anchorage was $4,684, while in Arizona it was $3,053/kW. Table 5.1 uses $3,873/kW, the average of these two locations, which is about the cost in Chicago ($3,849/kW) and Seattle ($3,697/kW). The PV system cost in a big part of the USA located south of the Chicago-Seattle line, which receives more solar energy, therefore would be cheaper.

Table 5.1 Electric power plants capital costs

		Nominal capacity (MW)	Capital cost ($/kW)	Cost of the system ($)*
Coal	Single unit	520	6,599	3,431,000,000
	Dual unit	1,300	2,934	*
Natural gas	Conventional	620	917	569,000,000
	Advanced	340	2,095	712,000,000
	Advanced*	500	2,095	1,048,000,000
Hydroelectric	Conventional	500	2,936	1,468,000,000
	Pumped storage	250	5,288	*
Uranium	Dual unit nuclear	2,234	5,530	*
Wind	Onshore	100	2,213	*
	Offshore	400	6,230	2,492,000,000
Biomass	Biomass combined cycle	20	8,180	*
PV	2012 price	20	4,183	*
	2012 price	150	3,873	*
	2013 price**	500	3,500	1,750,000,000
	2015 price**	500	2,800	1,400,000,000

*Only the cost of comparative capacity systems are listed.
**The author's estimate.

These indicate that a utility-size PV system's cost today is below that of a similar-size coal utility and is about the same as a hydroelectric system. It is important to note that neither the hydro nor the PV systems need to buy fuel while the gas and coal systems do. This means that the electricity produced by a PV system today is price wise very competitive with the electricity produced by coal, gas, and also hydro systems.

While the cost of building of a utility-scale PV system is in the lower end of the range of other systems, there still remains the insurmountable barrier: Establishing a utility-size PV power station also requires a large amount of capital. But the PV systems have two attributes whose adverse effect the utilities did not realize and anticipate. One, mentioned before, is that the PV systems need not be used only as a large "utility-scale" system. PV systems could be decentralized and mounted anywhere and in any smaller or larger size. The other, which the utilities did not consider, is that PV generates electricity during day time, and its highest electricity production is during the hours before and after noon when the peak demand of electricity is.

As already mentioned, one of PV's attributes is that it is a decentralized electricity-generating system. Table 5.2 indicates the installed cost of a small PV system on the roof of homes in two countries.

Table 5.2   Cost of an installed home system (2015)

Country	PV system size Roof mounted	Installed cost ($/kW)	Installed system cost ($)
PV—Germany[1]	10 kW	1,500	15,000
PV—USA[2]	10 kW	0	0

[1]Wolfgang Palz, Private communication, to be published.
[2]See Chapter 2.2.

In Germany, as it was shown in Chapter 2.2, by the end of 2011 about 20,000 MW of small, less than 1 MW size systems were installed on people's homes, of which 85% were roof mounted. The cost in Germany at present is $1,500/kW assuming they were 10 kW size systems on the homes. The owner in those days (2012) would have paid probably $25,000 (today $15,000). Compared to the value of the house, this is a small amount. Connecting that to the grid will make the owner of the house an IPP. Assuming that 2 million homeowners invest only $25,000 each produces a total of 20,000 MW electricity. This broke down the utilities' insurmountable barrier because these people suddenly jointly created a "utility" dumping electricity into the electric grid in an amount that is equivalent to the 24/7 operation of six 500 MW coal, gas, or nuclear power plants.

In the USA, as described in Chapter 2.2, there are companies that give the homeowner the option of installing the system on the roof of a home free of charge, and the homeowner will pay for the electricity used at a price that is below the rate the utility is selling it and fixed for 20 years. The owner also has the option to buy the PV system outright by either paying or lately more often taking a bank loan.

To make things even worse for the utilities, in the United States and also in Europe, the peak demand hours for electricity consumption are around noon and during the afternoon hours, especially during late spring, summer, and early fall months when the air conditioning load is high. As described in Chapter 1.2, the electric utilities had to install "peaking power plants." Because these "peakers" are used only when needed, they are inefficient investments; therefore the price of peak demand electricity is not regulated, and the utilities can charge as much as they want.

What the utilities did not anticipate is that the PV system's power production is during daytime, peaking around noon. For example, in Europe, by 2011 probably all of the grid-connected residential PV systems, totaling about 20,000 MW, were feeding electricity into the grid during peak hours. Therefore, the utilities' "peakers" were actually not needed, and the utilities had no income from them. The loss of this income was a very unpleasant surprise.

The major German electrical utilities expressed some interest in wind energy as that was a little bit closer to their business model. On the other hand, the utilities did not buy any PV systems as they were viewed as expensive technical curiosities, which they believed needed research to find a breakthrough to become inexpensive and useful. An example is E.ON, the second largest German utility, which was involved in little wind energy projects and not at all in PV. Dr. Johannes Teyssen, chief executive officer and a member of the management board of E.ON, noted in the company's 2011 annual report the existence of PV, which he calls "distributed generation": "Distributed generation also has the potential to play an important role in the energy world of the future. A team of experts drawn from across our company is currently designing a strategy for propelling E.ON's growth in this area."

Only nine months later at the end of the third quarter of 2012, in E.ON's third-quarter report, Dr. Teyssen realized that the future when the "distributed generation" would play an important role in the energy world must have been started a few years earlier, because he sounded the alarm bell as he realized that: "In most European markets, the gross margin for gas fired units is approaching zero or is indeed already negative. One factor is that the demand for electricity remains very low. But another key factor is that renewable-source electricity is being fed into the grid during peak load periods."

This observation was true not only in Germany but in many other countries, especially in the United States. When the price of electricity produced by PV systems approached the price the utilities were charging for non-"peak" hours, the difference in "peak" hours between PV electricity and the utilities price was considerable as described in Chapter 2.4. Some large department stores, e.g., Walmart, IKEA, Macy, COSTCO, etc., and many industrial organizations (e.g., Apple, Dow Chemical, Google, Amazon, L'Oreal, etc.) started to install PV systems realizing that they could save lots of money. Some of them declared to use 100% "green" electricity and installed large PV systems on the roof of their buildings and around their facilities (see Chapter 2.4). That resulted in more loss of revenues for the utilities, reducing the income.

Some of these industrial organizations are now going one step further and have started to compete directly with the utilities. Currently, when private companies sell their excess power, they can only do so to sell at wholesale rates. Apple, which uses clean energy, mostly PV, to power all of its corporate offices, data centers, and most of its Apple Stores, also formed a subsidiary Apple Energy LLC and applied to the US Federal Energy Regulatory Commission (FERC) to sell from the sites of its solar installations electric power all over the USA at market rates directly to end-users and get paid retail prices for its excess power rather than sell wholesale and get only wholesale prices. Google started "Google Energy LLC," and FERC on February 23, 2010, granted Google an authorization to buy and sell energy at market-based rate.[9] As a result of these examples, other companies or investors could create mini utilities and obtain from FERC the permission to compete with utilities. As mentioned, the utilities felt safe that the high cost of a power plant utilizing coal, gas, or nuclear fuel was an insurmountable barrier to prevent the competition from being established. The emergence of these mini utilities demonstrated that the establishment of an electric power plant does not require high cost and diminished the utilities' last insurmountable barrier protecting their monopoly.

In Germany the four major utilities, RWE, E.ON, EnBW, and Vattenfall, finally realized their problem, and they all turned suddenly green. They became "solarized." They created RWE Solar, E.ON Solar, EnBW Solar, and Vattenfall Solar. The RWE Solar Web site declared: "Switch now to smart energy. Make your roof a profit center." E.ON Solar: "You should be your own electricity producer." EnBW Solar: "Innovative all-round-carefree-package for Photovoltaic customers." Vattenfall Solar: "Without moving parts, noise and emission-free, thousands of plants in this country provide electric power." (The quotations are translations from the German Web sites of these companies). It is hard to believe, but they are all talking about PV. E.ON sells electricity not only in Germany but also in other countries in Europe, and they started to promote "solar" there too. For example, they also advertise their solar program in Hungary: http://www.eon.hu/solar/.

The situation the large German utilities were facing was very similar to what happened in the telephone business, which also started as a monopolistic business, regulated by localities and governments, with central stations connected to customers by wires. Same as utilities, it was a stable and profitable but regulated business. The emerging cell (mobile) phones, which were not connected to the local central station by wire but could be used anywhere in the world, became a fast-growing and, importantly, an unregulated business, where profit is limited only by competitors.

Telephone companies realized this during the nascent stage of the cell phone revolution. They had the vision to overcome their "corporate culture" of the wired phone business. They

---

[9]http://www.ferc.gov/whats-new/comm-meet/2010/021810/e-18.pdf.

established independent separate corporate structures for their unregulated cell phone business and put their money into it. They even unloaded the wired telephone business, which was not a growth area anymore. In May 2009, Verizon sold its wired telephone business in 12 mostly western states of the USA to Frontier Communications Corporation.

The two largest German utilities, E.ON and RWE, elected to do the same. E.ON, realizing in 2012 that PV is causing a serious problem, was the first to split up two years later in December 2014. The above-quoted CEO of E.ON, Johannes Teyssen, announced that the company would split into two separate, distinctly different entities. One would focus on renewables, distribution networks, and customer solutions. The other independent company ("New Company") would operate the conventional generation and global energy trading. This is exactly what the US telephone companies were doing. The distinction between these two independent companies is that E.ON (the one that is operating the RE segment) will be a growing and expanding profitable business, while "New Company," operating the conventional energy generation business, will be a stable low-growth business providing dividends for its investors.

RWE the other large German utility did exactly the same. A year later in December 2015, Peter Terium, the CEO of RWE, announced that RWE would also split into two separate companies, which de facto happened on April 1, 2016. Interestingly, Peter Terium would be the CEO of the "green RWE" and somebody else would be the head of the "conventional RWE." As of September 1, 2016, RWE's subsidiary was branded "innogy SE," standing for innovation and technology, which pools RWE's German and international renewables, grid and infrastructure and retail businesses.

"innogy SE," the renewables part of RWE, made its debut on the stock exchange on October 7, 2016. RWE received $3.35 billion for the shares. This demonstrated the success of RWE's decision to separate the "green" business from the "old" centralized and shackled utility model which would have difficulty to raise even a fraction of that money. Peter Terium, who was the CEO of one of the World's largest electric utilities, was elected to head its new subsidiary to become the CEO of "innogy," the "green RWE," and somebody else will be the head of the "conventional RWE." Peter

Terium defined the situation: *"Ten months ago, we set out with a very ambitious goal to create an innovative green energy company, which has the strength and independence to define the energy market of the future which is green, decentralized and digital".*

Other countries in Europe are also shifting to RE, wind, and PV. Chapter 4.4 provides an overview of the utilization of PV in several countries. As of European utilities outside of Germany the situation in France is very interesting, because Électricité de France S.A. (EDF; Electricity of France), the world's largest producer of electricity in 2011, produced 22% of the European Union's electricity. Most of EDF's electricity is produced by nuclear; therefore, they cannot shut down their nuclear power systems like Germany, but it is gradually adding more and more of the electricity produced by renewable energy, which at present is 12.3% (includes 4.6% hydroelectricity), almost as much as it produces from coal (14.5%) and more than from gas (8.6%).

In the USA the situation is more complicated. There are 3,269 utilities and most of these utilities still try to ignore the existence of PV even when globally already over 200,000 MW, and in the USA 35,375 MW solar PV generation,[10] is in operation. This 35,375 MW PV capacity is enough to power 7 million homes.

Recognition of the problem caused by PV systems for the utilities was described in the USA only in January 2013 by the Edison Electric Institute, which is the association of US investor-owned utilities (IOU), in a published report: "Disruptive Challenges: Financial Implications and Strategic Responses to a Changing Retail Electric Business."[11] This extremely interesting publication described the effect of PV on the electric utilities. It says that PV—a disruptive technology—is emerging and could directly threaten the centralized utility model.

In the USA, three types of utilities exist: investor-owned utilities (IOU), publicly owned utilities (owned by communities), and co-op utilities (co-ops are nonprofit utilities that are established to provide at-cost electric services to their member-customers). Of these groups the investor-owned utilities (188 exist) are the ones that started to take the utilization of PV electricity-generating systems more seriously. The top 10 IOUs utilize PV systems producing a total 3,582 MW. The top 10 of the

---

[10]https://www.eia.gov/electricity/monthly/epm_table_grapher.cfm?t=epmt_1_1.
[11]http://www.eei.org/ourissues/finance/Documents/disruptivechallenges.pdf.

public utilities utilize PV systems producing only 141 MW, while the top 10 of the co-op types are involved in only about 65 MW (2014 data). In many cases, the utilities were forced by a mandate of their local energy commission to provide electricity generated by PV. The lack of seriousness of the US electric utilities can be seen by comparing these amounts with the above-mentioned 35,375 MW PV capacity deployed in the USA.

But as of now, still only a few of the US utilities realize that PV systems do exist. For example, the Solar Electric Power Association (SEPA) in 2015 organized a "Utility Solar Conference." Only 75 utilities participated out of the over 3,000 US utilities. The US utilities are still fighting solar instead of realizing, like the major German utilities did, that traditional electricity generation is stagnating and the growth is in the PV systems.

However, the other main problem for the utilities, as mentioned before, is that more and more IPPs have emerged. Customers started to produce some or all of their peak time electricity by installing PV systems. These commercial installations were on office buildings and corporate campuses, large and small retail stores, manufacturing facilities, data centers, warehouses, and convention centers.

Furthermore, in the last years several IPPs developed large utility-scale PV fields. At least 15,000 MW of these systems are now in operation and 53,000 MW are under construction or under development.[12]

Because the utilities started to feel the erosion of their business, some of the IOUs started to develop strategies to compensate for this problem. They developed three different approaches. Few followed the example of the big German utilities. They separated the regulated part of their operation from the unregulated RE generating business. But none of them took the bold jump the Germans did. They did the separation, but they did not split up the regulated and the unregulated part of the business into separate corporate entities. They still have not admitted that the growth is in the unregulated business and the traditional is not a growth, but only a money-making machine to pay dividends.

Some of the utilities started to establish their own RE electricity-generating systems but did not separate that from the

---

[12]http://www.seia.org/research-resources/major-solar-projects-list.

conventional regulated business. More adapted the scheme that they may own some of the RE generating systems but will rely on power purchase agreements (PPA) to buy RE electricity from the IPPs. They were buying it from large utility-scale PV fields on a long-term contract. Some of them went as far as helping homeowners install PV on their roof and have a long-term power-buying agreement with them.

The majority of the utilities did nothing. They continued to fight the existence of the PV systems, but the present system is going to face more changes.

One is the development of the inexpensive electric storage system (discussed in Chapter 2.8) and the "smart controller." The German utilities, when they turned green, offered not only to install PV systems on the roof of homes but also to install batteries for the storage of electricity (discussed in Chapter 2.2). They started this in 2013.

The German utilities were smart. They lost regulated rate customers but gained in that these customers would not dump their PV electricity into the grid at the "peak power time." They made money installing PV systems, including storage batteries and the Bosch "smart controller" (Chapter 2.2), and retained the customer by selling them a long-term contract to maintain their PV systems.

In the USA, the majority of the utilities are trying to maintain the status quo: central power stations assuring investors with steady dividends, dealing with the various regulating agencies, providing electricity for customers utilizing cheap pole-mounted wiring, exposing them to power failures possibly lasting for weeks and to no price stability.

Unlike the German large utilities, the US ones did not realize that they are trying to stop a fast-moving train by putting matches on its tracks. They are standing on the platform and the train is running by. In 2010, the total PV installation in the USA was 2,000 MW (2 GW) and five years later in 2015, as mentioned, it was 35,375 MW (35.37 GW), close to 18 times more.

The private home PV market into which the large German utilities entered is serviced in the USA as described in Chapter 2.2 by startup companies with large capitalization. It appears US utilities have no intention to enter this market.

The utilities will lose the business of a very large number of homes that will be equipped with solar systems with electricity storage. Also, the utility-scale PV systems are IPPs; they could be bundled by some investor and will become a competitor to the utility in that area.

That the wheel is turning towards PV is evident in that in 2015, for the first time ever, more solar system capacity was added than natural gas. Solar supplied 29.4% of all new electric capacity in the USA.[13]

Another recently developing threat, especially for the smaller utilities, are the microgrids discussed in Chapter 2.3. They can provide higher reliability, affordability, safety, cost saving, and long-term price stability for housing developments and also for existing villages and sections of cities.

Considering all of the above, what is going to be the future of the US utilities? In 2013, it was very easy to make the prediction[14] of what will happen to the four large utilities supplying 67% of all electricity utilized in Germany (2013 data)— that they would split the company in two, conventional and new electricity production, and would put their money in the new "green" segment.

The crystal ball to look for prediction for the US companies is very murky as there are over 3,000 electric utilities, among them 188 large investor-owned ones. There are, however, a few things that help in making some predictions.

American utilities missed their opportunity to solarize homes, which would have been an easy market for them. They had the mailing list and they knew the pattern of electricity use in every home in their territory. Because of their stability, customers would have trusted the utility more than they would trust a newcomer offering to solarize their home. Utilities would get capital more easily than a new venture like SolarCity, Sungevity, Sunrun, Vivint Solar, or others. The large German utilities figured this out when they suddenly became "green." The window for the US utilities is still open for a short time, but it is not likely that they will do it.

The commercial installations and the utility-scale PV fields are the other area of PV systems causing losses for the utilities. Utilities can continue to do what they started to do, buying

---

[13]http://www.seia.org/research-resources/us-solar-market-insight.
[14]http://www.energypost.eu/future-large-german-utilities-already/.

electricity on long-term contracts from IPPs who established utility-scale PV fields or buying utility-scale fields for themselves. Obviously, they will use PV systems for underpinning their existing distribution systems, and they could establish electricity storage systems to support their "peaking" business.

If more and more utility-scale PV field owners will turn to FERC to get authorization to sell energy at market-based rate, then ultimately the electricity business will turn into a deregulated business similar to cable TV, cell phones, air transportation, etc. Utilities will be around for a long time, but their life will not be as easy as it was. PV started to cause major changes for them and created a new world for the electric utilities.

6

The Future of PV

Outlook to the Future

Wolfgang Palz[1]

It was an extraordinary achievement that PV became recently—against all odds—mainstream in global energy supply and consumption. From virtually nothing at the beginning of the new century, PV was spread worldwide to some 300 GW[2] of operational power capacity at the end of 2016. Most of it, some 250 GW, was installed in the last 6 years only. We are witnessing a political, financial, social, and industrial revolution.

The international political support started with the EEG legislation that became effective for PV in Germany in 2004 all the way to the federal Investment Tax Credit for PV in the United States, which was extended end of 2015 until 2023.

Global investments in PV from private sources crossed US$ 600 billion in total with an average of some $100 billion per annum over the last few years.

Over a million clean jobs in PV came about worldwide. Millions of individual PV generators are privately owned. They allow for the first-time "self-production "and "self-consumption" of on-site electricity generation. Zero-Energy and Plus-Energy Buildings are proliferating on a large scale thanks to PV. In support of the poor in the developing world, solar lighting becomes available at affordable cost thanks to PV in combination with the new LED technology.

Hundreds of new production, installation, and service companies have been created. The stock market value of the leading PV module manufacturers, most of them from China, taken together comes close to $10 billion. A star is the thin-film manufacturer First Solar in the USA at a market value of $4

---

[1]Wolfgang Palz's biography is on page 291.

[2]The PV capacity of 300 GW corresponds to a total solar module area of over 4000 km^2 exposed to the sun's irradiation. This area is a bit smaller than the land area of the small US State of Delaware. The PV area comes not in one block but is spread in millions of generators around the globe.

Sun towards High Noon: Solar Power Transforming Our Energy Future
Peter F. Varadi
Copyright © 2017 Peter F. Varadi
ISBN 978-981-4774-17-8 (Paperback), 978-1-315-19657-2 (eBook)
www.panstanford.com

billion. The world's leading large-scale regional or national electricity providers or utilities, such as EdF in France and E.ON and RWE in Germany, switched for economic reasons—not that they suddenly became lovers of solar energy—strategy away from their traditional business based on the conventional sources of energy to PV and the other renewables.

All countries around the world are currently already benefiting from PV. Some 25 nations operate at this moment PV capacities in the GW range, including many developing countries. World leaders are China, the United States, Japan, and Germany with respectively over 30 GW in operation at the end of 2016. Currently, an exception to the global growth of PV deployment seems to be Russia: The largest PV capacity it owns lies in Crimea, which the Ukrainians had developed there.

Besides China, the United States has recently become a key promoter of PV. The PV market in the USA turned red hot in 2016 doubling new capacity installation with respect to 2015. According to the US Energy Information Administration (EIA), new PV surpasses all other power capacity installations in 2016, beating that of new coal and natural gas. It is amazing for a country like the USA that leads traditionally the world in coal consumption—right after China. For the first time, new PV exceeds even new wind power, the other frontrunner in renewable electricity. The USA is catching up with China as the world's brilliant second in PV deployment.

PV has turned from an ecological dream to a commercial success. In 2016 PV solar electricity became available at costs as low as 3 US cents/kWh ($30/MWh). This was achieved for large multi-megawatt PV installations in the "solar belt" of the globe. Even in poorer solar climates, costs down to 6 cents/kWh are now possible.

Such low kWh costs of PV electricity became possible by two effects. One is the low lending rate of capital. Bank rates have currently come close to zero in the Western world. That is an encouragement for investments looking for profitable opportunities. The other effect comes with the low kW system costs of PV that were achieved over the last few years. Leaving alone the conventional fossil and nuclear plants that are amortized, the new reality in the electric power business is that the PV of today has become competitive with new coal, gas, and nuclear plants. Large-scale PV plants are currently built at $1,000/kW—

higher costs up to twice that value and more are experienced, too, but they apply for smaller PV systems and local markets that are still in their infancy and learning phase.

The US EIA has published "cost estimates for generic utility scale generating plants from 2012." New advanced coal plants are listed at \$4,400/kW, new advanced gas plants at \$1,000/kW, and nuclear at \$5,500/kW. One might argue that PV plants have lower operation time—even in the solar belt PV operates less than 2000 hours of equivalent full power in a year. But contrary to fossil and nuclear plants, PV does not need any other fuel than the sun and O&M is a lot lower. And PV does not have those harmful greenhouse gas (GHG) and waste emissions—and it is not a nightmare to dismantle the plants at the end of their operational life.

The heart of a PV generator is the solar module. In a well-established market, the cost of the module makes up approximately half of the total system cost. The rest of the cost goes for the BOS, installation, administrative, and other service costs. Module costs came all the way down to \$0.5/W. In the mid-1970s, some 40 years ago when working in Paris, I had offers for solar modules from Solarex in the US at \$10/W. That was the best offer one could get in those days. A cost reduction of a factor 20 has been achieved, not even including inflation since those days!

In terms of solar cell and module costs, there was a race since the early days in the 1960s between silicon and the family of thin-film semiconductors such as CdS, CIS, CIGS, a-silicon, micro-crystalline silicon, and GaAs for optical concentration. Experts expected the TF at the goal post, but the winner was the classical silicon. Relying on the basic silicon technologies developed in the 1970s—the blue cell, the violet cell, or the black solar cell—the silicon industry achieved what not many had expected, that cost of 50 cents/Watt. That is the more striking as an army of thousands of researchers at universities and specialized solar institutes attempted over all these years to upgrade the old module technologies. Following RTS,[3] crystalline silicon provided 93.4% of global PV module production in 2015, TF CdTe 4%, TF CIGS 1.6%, and TF Si less than 1%.

---

[3]*PV Activities in Japan and Global PV Highlights*, published monthly by RTS Corporation, Tokyo 104-0032, Japan, www.rts-pv.com.

The other surprise in the new PV world was the emergence of the Chinese module industry. China was an absolute newcomer when Germany opened first the new industrial era of PV deployment in 2004. Until that year, China never had a PV industry and only an insignificant PV research community. And ultimately China's industry won with flying colors the race towards the 50 cents/watt module—at the expense of the silicon cell industry in Germany, Japan, and other countries. A real thriller. Following RTS, China and Taiwan produced in 2015 50 GW of solar cells, the overwhelming share of the global cell production of 63 GW that year. And 70% of all PV modules produced worldwide came from China. Of the 44.2 GW of PV modules produced, China installed 15.15 GW within the country itself and the rest was exported. Not only was China in 2015 the world's largest module producer, it also had the largest domestic market.

How about the future of PV in the global markets? From what we have seen before, it is easy to speculate on the short-term outlook of PV by simple extrapolation. Societies have by now learned about the benefits of PV, technologies are well established, and the world has a solid PV industry that learned from the past and got rid of the ups and downs it went through.

In the last few years, the global PV markets were always on the rise and there is no reason why this would change in future. Growth rates are expected to become more moderate, however, as we are already on a high market volume of yearly installations.

Extrapolating from the 50 GW of PV installed globally in 2015, for the 4 years from 2016 to 2020, PV markets can be conservatively expected to rise by between 60 and 70 GW a year worldwide. By 2020 then a total of some 500 GW and more would be in operation globally. China has already announced that it aims for 143 GW in operation by then. In terms of investments, a total of $1 trillion ($1000 billion, or 1 million of a million USD) will have been spent on PV deployment since the beginning of the market rush in modern times.

For the years beyond, the outlook becomes more uncertain. Assuming that yearly installation rate would climb up to 100 GW per year on average between 2020 and 2030, eventually 1,500 GW would have been achieved. The International Renewable Energy Agency announced in June 2016 a PV capacity volume of between

1,760 GW and 2,500 GW in operation in 2030. Maybe, but how can they know as PV is dependent on random and unpredictable political decisions in the future?

Let us address at long last also the long-term perspectives of PV in the world energy markets. What could ultimately be its contribution to a world relying again on a 100% renewable energy supply—in case it is realistic to expect it? For this, we have to go back to some fundamentals: on the one hand, the inconveniences PV may face when applied on a very large scale, and on the other hand, the promotional instruments as part of the global policy in favor of the promotion of all the renewable energies.

Let's first consider some deficiencies of PV. As it follows, the sun's radiation and its time of generation over the year is one of the most limited among all power options and its availability is intermittent. Also, PV is an electricity producer and as such cannot meet the world's energy demand for heat and transport.

When PV capacity of 500 GW might have been reached in 2020, that would mean that already some 7% of the overall power capacity of some 7,000 GW of all sources installed around the world by then would consist of solar PV—not a minor achievement in just one decade. But as PV's operational time over the year is relatively limited—it is sub-average in the overall power system—its contribution to the global electricity generation will only be 2% of the expected 26,000 TWh of total consumption. Assuming that in the more distant future, at one stage PV might contribute 20% of the world's electricity supply, that would mean a PV capacity installation of 5,000 GW, i.e., more than half the power from all sources together.

And because of its intermittency, solar electricity would at best be available 15% of the time at full equivalent power. Solving the intermittency with storage devices is possible but not a cheap idea. In particular, seasonal storage to bridge the solar availability between summer and winter would not be cost optimal: It is like buying a car that is used only a few times a year.

In view of the vision of a clean world without GHG emissions and a 100% energy supplied by renewable sources, the possible role of PV faces new hurdles. Today, 40% of the world energy consumption is for electricity, but heating and transport comprise 40% and 20%, respectively. It does not seem very likely that PV will have much impact on the latter sectors. In conclusion, even at 5,000 GW installed, PV would contribute only some 7% of the overall global demand.

There is no doubt that PV is an excellent candidate to benefit from the political promotion of all the renewable energies (RE) of which it is a part. The world suffers increasingly from pollution and climate change and the clean REs are—together with energy conservation—the only way out of this disaster for humanity. A new step forward in this direction was the UN Climate Conference COP21 that took place at the end of 2015 in Paris, France. It was of extraordinary political importance with the personal attendance of the entire world's leading heads of state. The treaty adopted at the meeting and signed and ratified in the meantime by many countries demands a global limitation of temperature rise as an effect of global change to 2°C. Correspondingly, it has become imperative to eliminate burning of fossil fuels and switch to a maximum consumption of RE instead.

Evidence was presented previously that PV alone is not suitable for leading a change to a cleaner world. It would be unrealistic because of the limitations of PV. But PV should eventually lead a symphony of all RE sources together. PV is respected everywhere in society. It has an advantage over the other RE sources, which encounter various degrees of opposition from conservative quarters, some greens and ecologists, or both. It is the case for hydro, in particular large hydro dams, for wind turbines, for biomass transport fuels, and for other heating applications, for geothermal, etc.

There is opposition, but that does not mean that it cannot be overcome. From a practical point, the combination of the use of PV together with RE sources such as hydro and biomass that are not intermittent looks especially attractive in the long term, when the classical fuels will be marginalized.

The debate on the long-term perspectives of the renewables is increasingly focusing on the need for a 100% RE world. It is a grassroots movement from all parts of society. Would it be realistic to eliminate all conventional energies one day and replace them with the renewables? If yes, by when?

First, the debate concerns conventional oil, which is currently the world's leading energy. Obviously, it will not be replaced very easily. In practice, it is attractive because of its convenience in many applications, in particular transportation. The price collapse of oil and natural gas in the international markets in 2016 was a nightmare for the economy of the exporting countries such as Venezuela and Russia. Despite strong declarations of some producers such as Saudi Arabia to join those important quarters in society to abandon oil and natural gas production and use, it looks doubtful that oil and gas could one day be abandoned completely in the energy markets.

At first glance, things look more favorable for the RE in the electricity sector. As over 50% of new power generation installations in the world since 2015 were the RE type, there is a clear trend towards the domination of the new clean power in the electricity world. However, it must not be overlooked that most of the electricity we consume today stems from the conventional sources. The switch to a 100% RE world is not easy. Most of the conventional power plants have a long service life of 50 years and more. And most of them are amortized and produce electricity very cheaply. It will be a Hercules task to get rid of the existing and well-operating 4,000 GW of fossil and nuclear generation plants worth many trillions of USD. The most difficult to eliminate will be the hundreds of atomic plants now in operation. Take the 60 atomic plants that provide most of the electricity consumed in France. Eliminating them quickly would mean a collapse of its economy. Consequently, the French government, which owns them, is not aiming to replace them but to upgrade them to extend their operational life. And the world has not to deal only with the existing ones; there are also a lot of new ones on the drawing board: China, for instance, intends to build 139 GW of new nuclear plants by 2040.

It is not much better for fossil energies. In June 2016, the energy chiefs of the G20 nations meeting in Beijing could not agree on a deadline to phase out fossil fuel subsidies. Those are estimated at over $440 billion a year. And in Germany, once a pioneer for RE deployment, the government decided in 2016 not to reduce its effort in the coal sector before 2050.

In conclusion, the progression of the renewables in the world energy markets is unstoppable. The move is carried by vital parts of society. But the task is enormous and often underestimated. A complete switch to a real solar energy world is not for tomorrow.

As far as PV is concerned, it will continue its own path of success. It is cost competitive with most other sources of electricity and has a strong support in society. The door is wide open for achieving thousands of gigawatts of clean and reliable power.

PV Power for the People—PV Power for the World.

About the Contributors

Series Editor

Wolfgang Palz has been continuously involved in the development of global photovoltaics (PV) as a scientist and manager for more than 50 years. He is currently the promoter of auto-consumption for PV for providing worldwide access of cheap electricity to all. He has been the leader of PV development in France since the 1970s, when the country was the European leader in the field. As an official of the European Commission in Brussels, he was the manager of PV development in Europe for 20 years. He inspired and supported German initiatives to kick off PV markets, which led to a global explosion of PV investments since 2004. Besides his key role in the development of PV in Europe, Dr. Palz is much connected to the United States, where he worked with NASA and is currently a member of the leadership council of ACORE in Washington, DC. He is equally active in China, where he recently organized the Green Wall Forum on Renewable Energies.

Contributors

Michael Eckhart is Managing Director and Global Head of Environmental Finance at Citigroup, Inc. He led Citi's work in establishing the Green Bond Principles and is actively engaged with REN 21, the IEA's Renewable Advisory Board, the International Renewable Energy Agency (IRENA), and the Atlantic Council. Previously, he was founder and President of the American Council on Renewable Energy (ACORE) in Washington, DC, and the SolarBank Initiative in Europe, China, India, and South Africa. Previously, he had a 20-year career in power generation with United Power Systems, Aretê Ventures, General Electric, and Booz, Allen & Hamilton. He has received numerous awards, including Renewable Energy Man of the Year of India in 1998, the Skoll Award for Social

Entrepreneurship in 2008, and ISES's Global Policy Leadership Award in 2013. He served in the US Navy Submarine Service and holds engineering and business degrees from Purdue University and Harvard Business School.

Allan R. Hoffman holds a Bachelor of Engineering Physics degree from Cornell University and a PhD in physics from Brown University. Trained as an experimental physicist, he has devoted most of his career to the planning and management of clean energy technology programs in Washington, DC. He has served as Staff Scientist for the US Senate Committee on Commerce, Science, and Transportation, in senior positions at the National Academy of Sciences and the US Department of Energy, and as Vice Chairman of the International Energy Agency's Working Party on Renewable Energy. Dr. Hoffman is a Fellow of the American Physical Society and the American Association for the Advancement of Science. He is the author of the book *The U.S. Government and Renewable Energy: A Winding Road.*

Paula Mints is the founder and Chief Market Research Analyst of the global solar market research firm SPV Market Research. Ms. Mints began her solar market research career in 1997 with Strategies Unlimited. In 2005 she left Strategies Unlimited for Navigant, where she continued her practice as a Director in Navigant's Energy Practice until October 2012, when she founded SPV Market Research. Ms. Mints published her first book, *Legacy of Courage*, in 2000. She is the author of many articles on the economics and behavior of the solar industry, specifically photovoltaic technologies and markets. Ms. Mints' work was cited in the US Department of Energy's current SunShot report, Photovoltaic System Pricing Trends, 2014 Edition. Her chapter, Overview of Photovoltaic Production, Markets and Perspectives, was published in the Fraunhofer/Elsevier book *Advances in Photovoltaics Volume 1*, edited by Dr. Gerhard Willeke and Dr. Eicke Weber. Ms. Mints speaks at several conferences annually, including Intersolar North America, SPI, the IEEE PVSC, and the

EU PVSEC and is on the expert committee for the EU Photovoltaic Technology Platform (www.eupvplatform.org). Ms. Mints earned her MBA at San Jose State University in 1999.

Bill Rever is the co-founder and Chief Marketing & Sales Officer of Advanced Silicon Group (ASG) and a consultant in solar and renewable energy. He began his PV career in 1982 with industry pioneer Solarex, which later merged with BP Solar. Bill held a variety of roles within that company culminating as Strategy Director. In his tenure, Bill was involved in the production, marketing, and deployment of over 1 GW of PV in over 150 countries, including many seminal applications, product innovations, and projects. Bill has a BA in Physics from the Johns Hopkins University, an MSE in Energy Engineering from the University of Pennsylvania, and an MBA from that University's Wharton School of Business. He is a member of SEIA, MDV-SEIA, and ASES. He is a former board member and past President of the Maryland/DC/Virginia chapter of SEIA and former Co-Chairman of the PV Advisory Group of NA SEMI.

John Wohlgemuth joined the National Renewable Energy Laboratory as Principle Scientist in PV Reliability in 2010. He is responsible for establishing and conducting research programs to improve the reliability and safety of PV modules. Before joining NREL he worked at Solarex/BP Solar for more than 30 years. His PV experience includes cell processing and modeling, Si casting, module materials and reliability, and PV performance and standards. Dr. Wohlgemuth has been an active member of working group 2 (WG2), the module working group within TC-82, the IEC Technical Committee on PV since 1986 and has been convener of the group for more than 15 years. Dr. Wohlgemuth is a member of the Steering Committee for the PV Module QA Task orce (PVQAT) and he chairs Task Group 3 on Humidity, Temperature and Voltage. Dr. John Wohlgemuth earned a PhD in Solid State Physics from Rensselaer Polytechnic Institute.

 **Frank P. H. Wouters** possesses 25 years of international experience in the field of sustainable energy. From 2009 to 2012 he served as the Director of Masdar Clean Energy, where he was responsible for renewable energy projects representing enterprise value of more than $3 billion in Asia, Africa, and Europe. Frank Wouters was appointed Deputy Director-General of the International Renewable Energy Agency (IRENA) in September 2012, a position he held for two years. He currently leads the EU GCC Clean Energy Network in Abu Dhabi. Mr. Wouters has worked throughout his career with a wide variety of stakeholders, including the private sector and government officials at the highest levels. He has supported sustainable energy policy in many countries, including Abu Dhabi and Nigeria, and is currently advising Dubai's government on the renewable energy aspects of the Expo2020 site. Mr. Wouters holds a master's degree in Mechanical Engineering from Delft University of Technology, the Netherlands.

Index

Sun towards High Noon
Solar Power Transforming Our Energy Future
Peter F. Varadi
9789814774178 (Paperback), 9781315196572 (eBook)
2017

Sequel to

Sun above the Horizon
Meteoric Rise of the Solar Industry
Peter F. Varadi
9789814463805 (Hardcover), 9789814613293 (Paperback), 9789814463812 (eBook)
2014

Sun towards High Noon describes the development of three new gigantic PV markets and new financing methods, which resulted in an extraordinary upswing starting in 2011. As a result of the new markets and the new financial methods, PV's worldwide operational power capacity grew to 300 GW, out of which 250 GW was installed between the years 2011 and 2016.

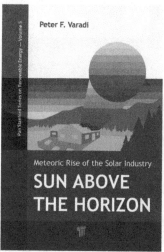

This explosive utilization of PV was based on the meteoric expansion of PV utilization during the years 1973–2010, which resulted in an incredible price reduction—described in the author's earlier book, *Sun above the Horizon*, a history of the terrestrial PV industry starting at the infancy when the first two terrestrial PV companies (Solarex and Solar Power Corporation) were formed in the United States in 1973, and when the "PV industry" employed 20 people and produced only 500 watts of PV power. The book guides the reader on PV's 40-year-long winding road to the time when mass production finally happened and incredible price reduction was achieved, which opened the doors to the extraordinary upswing described in this book.

Sun above the Horizon is a must-read, as can be seen from the following citations:

In *The Wall Street Journal*'s August 22–23, 2015, issue, Daniel Yergin writes in his review titled "Power Up": "Solar is growing fantastically," says Dr. Varadi, who chronicles solar's rise in his new book, *Sun above the Horizon*. "Something like this requires time. Shale oil and shale gas had a ready market. When we started, we had no market at all, zero. And the industry had to get to mass production to bring down cost."

Deloitte, the multinational professional services firm, announced their book selection for 2016: "Our featured book is *Sun above the Horizon: Meteoric Rise of the Solar Industry*, 44th book of the Books with Branko program."

Printed in the United States
by Baker & Taylor Publisher Services